FACTOR ANALYSIS AS A STATISTICAL METHOD

D. N. LAWLEY, M.A., D.Sc.

Department of Statistics, University of Edinburgh

and

A. E. MAXWELL, M.A., Ph.D.

Institute of Psychiatry, University of London

LONDON BUTTERWORTHS

THE BUTTERWORTH GROUP

ENGLAND

Butterworth & Co (Publishers) Ltd
London: 88 Kingsway WC2B 6AB

AUSTRALIA

Butterworth & Co (Australia) Ltd
Sydney: 586 Pacific Highway Chatswood, NSW 2067
Melbourne: 343 Little Collins Street, 3000
Brisbane: 240 Queen Street, 4000

CANADA

Butterworth & Co (Canada) Ltd
Toronto: 14 Curity Avenue, 374

NEW ZEALAND

Butterworth & Co (New Zealand) Ltd
Wellington: 26-28 Waring Taylor Street, 1
Auckland: 35 High Street, 1

SOUTH AFRICA

Butterworth & Co (South Africa) (Pty) Ltd
Durban: 152-154 Gale Street

First published 1963

Second impression 1967

Second edition 1971

ISBN 0 408 70152 8

Suggested UDC No 519-241-6

Printed in Hungary

CONTENTS

PREFACE

We are indebted to Messrs. Butterworth for having invited us to write the original version of this book. In doing so we had two aims primarily in mind. The first was to present the statistical theory of factor analysis, as far as it had then been developed, in a concise form that would enable statisticians to discover what the subject was about. The second aim was to illustrate the theory by some carefully chosen examples so that the ever-increasing number of research workers who used factor analysis would be enabled to do so in a more rigorous way than had formerly been the case.

When the first edition was written, the outstanding difficulty of applying the maximum likelihood method of estimation to factor analysis was the lack of satisfactory methods of obtaining numerical solutions. In the intervening period this difficulty has been almost entirely removed. The first break-through came with a paper by K. G. Jöreskog, published in 1966, in which he used a numerical method for the minimisation of a function of many variables that was superior to any that had previously been employed in factor analysis. Still better methods were later developed by Jöreskog in collaboration with others. It was largely as a consequence of these developments, of which a full account is given in the text, that a second edition of the book became necessary. In presentation and notation this edition differs radically from the first.

One important addition is an account of the sampling theory of estimates. In the first edition we concluded somewhat pessimistically that general formulae for the standard errors of loadings, even if they could be found, would prove to be too complicated for practical use. This conclusion was fortunately incorrect. In all cases expressions for the large-sample variances and covariances have now been found and, though by no means simple, they can readily be evaluated with the aid of an electronic computer.

The second edition contains two appendices. The first, on matrix algebra, is an enlarged version of the original one that includes, among other things, a section on vector and matrix differentiation. The second, on methods of minimising functions, has been added for the benefit of those readers who are interested in the details of the procedures for obtaining maximum likelihood estimates.

We are very grateful for the help that we have received from various quarters and, in particular, we express our thanks to Mr. P. O. White, who adapted Jöreskog's programs for use on the CDC 6600 computer of the University of London and who supplied several sets of numerical results used in the text. We also thank Mr. B. S. Everitt, who wrote the program for varimax rotation, and Mr. M. R. B. Clarke for making available to us results that he obtained by use of the Newton–Raphson method of function minimisation and for calculating the standard errors of *Tables 5.2* and *5.3*. Finally our thanks are due to Miss Nina Aitken who typed a difficult manuscript with exceptional care.

Scientists often come under censure for their misuse of language, and any literary merit that the book may possess should be largely attributed to our wives.

D. N. L.
A. E. M.

Chapter 1

THE SCOPE OF FACTOR ANALYSIS

1.1 INTRODUCTION

Factor analysis is a branch of multivariate analysis that is concerned with the internal relationships of a set of variates. Initially it was developed mainly by psychologists, with Spearman, Thomson, Thurstone and Burt as the most prominent pioneers, and was primarily concerned with hypotheses about the organisation of mental ability suggested by the examination of correlation or covariance matrices for sets of cognitive test variates. The early work in this difficult field gave rise to protracted controversies on the psychological side that for a long time discouraged the interest shown by mathematicians in the theoretical problems involved, and the subject became the black sheep of statistical theory. Gradually, however, a more coherent treatment of the subject has been developed, and now that many of the earlier misconceptions have been cleared up there is some hope that it may acquire a greater respectability. Today factor analysis is the most widely used of multivariate techniques, though not always appropriately, as Hotelling (1957) has pointed out. Its use has been greatly facilitated by the advent of electronic computers and is spreading to disciplines other than psychology, such as botany, biology, economics and the social sciences. An attempt to set out the statistical theory of the subject is clearly desirable.

In this introductory chapter the problems with which factor analysis deals and the models that it employs will be surveyed in a general way, while in succeeding chapters specific topics will be elaborated. Due to limitations of space, the numerical examples used in this book have had to be kept small and so are not altogether typical of factor studies, in which correlation matrices for forty or more variates are not uncommon, but they are adequate to illustrate the theory. The examples are also mainly psychological, as most of the published work is in this field. The general principles of factor analysis can nevertheless be readily applied in other fields.

1.2 FACTOR MODELS

For analysing the structure of covariance or correlation matrices two methods, that formally resemble each other but have rather different aims, are currently in use. One is principal component analysis, developed by Pearson (1901) and Hotelling (1933), while the other is factor analysis, which originated with the work of Spearman (1904, 1926). In the interest of clarity it is advisable to distinguish between the two approaches and though this book is primarily concerned with factor analysis the principal component method will also be discussed.

In principal component analysis, described fully in Chapter 3, a set of p variates, denoted by $x_1, \ldots, x_p$, is transformed linearly and orthogonally into an equal number of new variates $y_1, \ldots, y_p$ that have the property of being uncorrelated. These are chosen such that y_1 has maximum variance, y_2 has maximum variance subject to being uncorrelated with y_1, and so on. The transformation is obtained by finding the latent roots and vectors (see Appendix I) of either the covariance or the correlation matrix. The latent roots, arranged in descending order of magnitude, are equal to the variances of the corresponding y-variates, which are the unstandardised principal components. Often the first few components account for a large proportion of the total variance of the x-variates and may then, for certain purposes, be used to summarise the original data. All components are, however, needed to reproduce accurately the correlation coefficients between the x-variates. Hence the method is not appropriate for investigating their correlation structure. When it is employed no hypothesis need be made about the x_i. They need not even be random variates, though in practice this is usually the case.

In contrast to the method of principal components, the aim of factor analysis is to account for the covariances of the observed variates in terms of a much smaller number of hypothetical variates, or factors. Put simply in correlational terms, as by Howe (1955), the first question that arises is whether any correlation exists, that is whether the correlation matrix differs from the unit matrix. If there is correlation, the next question is whether there is a random variate f_1 such that all partial correlation coefficients between the x-variates after eliminating the effect of f_1 are zero.

If not, then two random variates f_1 and f_2 are postulated and the partial correlation coefficients after eliminating f_1 and f_2 are examined. The process continues until all partial correlations between the x-variates are zero. It is clear from this that whereas principal component analysis is variance-orientated, factor analysis is covariance- or correlation-orientated.

The aims of the two methods can also be contrasted by considering the nature of the relationships involved. In component analysis the y-variates are by definition linear functions of the x-variates and no question of a hypothesis arises. In factor analysis, on the other hand, the basic assumption is that

$$x_i = \sum_{r=1}^{k} \lambda_{ir} f_r + e_i \qquad (i = 1, \ldots, p), \tag{1.1}$$

where f_r is the rth common factor, the number k of such factors being specified, and where e_i is a residual representing sources of variation affecting only the variate x_i.

In equations (1.1) the p random variates e_i are assumed to be independent of one another and of the k factors f_r. The latter may be either correlated (oblique) or uncorrelated (orthogonal). Usually they are scaled to have unit variances. For convenience, and without loss of generality, we suppose that the means of all variates are zero. The variance of e_i, termed either the residual variance or the unique variance of x_i, is denoted by ψ_i. The coefficient λ_{ir} is known as the loading of x_i on f_r or, alternatively, as the loading of f_r in x_i.

In practice the λ_{ir} and the ψ_i are usually unknown parameters that require estimation from experimental data. In a paper by Whittle (1953) the f_r are also treated as parameters, instead of as variates. However, this raises difficulties when questions of sampling and estimation are considered, a serious one being that when the sample size tends to infinity so also does the number of parameters. Hence we have not followed this approach in our treatment of the subject.

It is clear that equations (1.1) are not capable of direct verification since the p observed variates x_i are expressed in terms of $p+k$ other variates that are unobservable. However, as we shall see in Chapter 2, the equations imply a hypothesis that can be tested regarding the variances and covariances of the x_i.

An important point, to which Bartlett (1953) has drawn attention, is that factor analysis involves a hypothesis of linearity. Though this might be expected to work as a first approximation even if it were untrue, it would lead us to reject the linear model of (1.1) if the evidence demanded it. Since correlation is essentially concerned with linear relationships it is not capable of dealing with this point, but Bartlett briefly indicates how the basic equations could be amended to give a better approximation. In their amended form, product terms would arise and with two factors, for example, the original equation

$$x_i = \lambda_{i1} f_1 + \lambda_{i2} f_2 + e_i \tag{1.2}$$

would be replaced by

$$x_i = \lambda_{i1} f_1 + \lambda_{i2} f_2 + \lambda_{i3} f_3 + e_i, \tag{1.3}$$

where $f_3 = f_1 f_2$.

The topic of non-linearity has been discussed more recently. See, for example, two papers by McDonald (1967a, 1967b) and other papers referred to therein. It appears that non-linear factor analysis has encountered some theoretical and practical difficulties and that it is not always easy to avoid ambiguities in formulating models. In discussing that of (1.3) McDonald concludes that much further experience with experimental data will be required before it can be decided whether or not the model is necessary or useful.

1.3 CHOOSING AND FITTING A SUITABLE MODEL

Psychologists have had differences of opinion as to whether factors should be conceived as correlated or uncorrelated variables. In the former case the correlation coefficients between factors are additional parameters that as a rule require estimation. A more serious cause of controversy in the past has been the fact that, with more than one factor, equations (1.1) do not by themselves determine either the factors or the parameters uniquely. For if the factors f_r are uncorrelated, they may be replaced by any orthogonal transformation of themselves, with a corresponding transformation of the loadings, while with correlated factors any non-singular linear transformation may be made.

In Chapters 2 and 6 we discuss various methods of removing this indeterminacy. The method that is appropriate in any given circumstances depends not only on what factors the experimenter expects to be present but also on what kind of hypothesis, if any, he is able to make concerning the data. In each case some restrictions must be placed on the parameters to ensure that they are uniquely determined. The general question of factor rotation and transformation is also discussed.

Having decided precisely what type of model is to be used, the next task is to estimate as well as possible the various parameters in it. For this purpose we have used the method of maximum likelihood. As a result it is then a simple matter to test whether the chosen model adequately fits the data. By contrast, the sampling theory of the various maximum likelihood estimators is somewhat complicated. It is now, however, possible to find large-sample formulae for their variances and covariances. This enables approximate confidence limits for the parameters to be found. An account of these topics is given in Chapters 4, 5 and 7.

1.4 FACTOR SCORES

In a principal component analysis the components are, as we have seen, linear functions of the original variates from which they have been derived. Hence there is no difficulty in estimating the scores of any individual on the components. In factor analysis, on the other hand, where the common factors do not fully account for the total variance of the variates, the problem is more difficult. Here the factors are not linear functions of the variates alone and the scores of an individual on them cannot therefore be found exactly. They cannot even be estimated in the usual statistical sense, and some minimum variance or 'least squares' principle has to be invoked in order that reasonable estimates may be obtained.

Estimation in factor analysis is thus, in a sense, a two-stage procedure. First the parameters in the model are estimated, then these are used to provide estimates of individual factor scores. In Chapter 8 we describe two methods of estimating factor scores, having slightly different properties. In each case the estimates are linear functions of the original x-variates.

1.5 COMPARING RESULTS FROM DIFFERENT SOURCES

A question that naturally arises in factor analysis is whether, for a given set of variates, the same factors occur in different populations. The difficulty in discussing this question is that, though equations (1.1) are to be regarded as invariant, the factors may undergo varying degrees of selection, so that their covariance matrices may in consequence differ from one population to another. Past attempts at dealing with this difficulty have not proved altogether successful and this has led us in Chapter 9 to suggest a new method for identifying factors in different populations. We also mention very briefly certain problems that may arise when estimates of factor scores are required for individuals drawn from more than one population.

Chapter 2

PARAMETERS IN FACTOR MODELS

2.1 INTRODUCTION

We shall now consider in more detail the factor analysis model represented by equations (1.1) of Chapter 1 and by the assumptions that follow them. It will be convenient to rewrite these equations in matrix form. Let $\mathbf{x}$ and $\mathbf{e}$ denote column vectors with respective elements x_i and e_i $(i = 1, \ldots, p)$ and let $\mathbf{f}$ be the column vector with elements f_r $(r = 1, \ldots, k)$. Then the equations become

$$\mathbf{x} = \mathbf{\Lambda f} + \mathbf{e}, \tag{2.1}$$

where $\mathbf{\Lambda} = [\lambda_{ir}]$ is the $p \times k$ matrix of loadings.

The covariance (or dispersion) matrix of the variates in $\mathbf{x}$ is defined as $E(\mathbf{xx}')$ and is denoted by $\mathbf{\Sigma} = [\sigma_{ij}]$, where σ_{ii} is the variance of x_i and where σ_{ij} is the covariance of x_i and x_j. This matrix is symmetric, positive definite and of order p. In view of the assumptions made about $\mathbf{f}$ and $\mathbf{e}$, we have

$$E(\mathbf{fe}') = \mathbf{0}.$$

For the present we shall assume that the factors are uncorrelated, with unit variances. The covariance matrices of $\mathbf{f}$ and $\mathbf{e}$ are then given respectively by

$$E(\mathbf{ff}') = \mathbf{I}_k,$$

where $\mathbf{I}_k$ is the unit matrix of order k, and

$$E(\mathbf{ee}') = \mathbf{\Psi},$$

where $\mathbf{\Psi}$ is a matrix whose diagonal elements are $\psi_1, \ldots, \psi_p$ and whose non-diagonal elements are zero. Since

$$E(\mathbf{xx}') = E[(\mathbf{\Lambda f}+\mathbf{e})(\mathbf{\Lambda f}+\mathbf{e})'],$$

we have

$$\mathbf{\Sigma} = \mathbf{\Lambda\Lambda}' + \mathbf{\Psi}. \tag{2.2}$$

In practice the elements of $\mathbf{\Lambda}$ and $\mathbf{\Psi}$ are unknown parameters that have to be estimated from experimental data. Problems of sampling and estimation will be considered in Chapters 4, 5 and 7. Here we shall concentrate solely on questions concerning the parameters. In particular we shall discuss the problems of defining them uniquely. Given the matrix $\mathbf{\Sigma}$ we ask whether for a specified value of k, less than p, it is possible to define a unique $\mathbf{\Psi}$, with positive diagonal elements, and a unique $p \times k$ matrix $\mathbf{\Lambda}$ satisfying equation (2.2). Unless this can be done, consistent estimators do not exist.

2.2 ROTATION OF FACTORS

Let us first suppose that there is a unique $\mathbf{\Psi}$. Then the matrix $\mathbf{\Sigma}-\mathbf{\Psi}$ must be of rank k. This is the covariance matrix for $\mathbf{x}$ in which each diagonal element represents not the total variance of the corresponding variate but only that part which is due to the k common factors; this is known as the 'communality' of the variate.

If $k = 1$, then $\mathbf{\Lambda}$ reduces to a column vector of p elements. It is unique, apart from a possible change of sign of all its elements, which corresponds merely to changing the sign of the factor. Such sign changes are merely trivial and we shall ignore them when discussing the uniqueness of vectors.

In cases where $k > 1$ there is an infinity of choices for $\mathbf{\Lambda}$. For equations (2.1) and (2.2) are still satisfied if we replace $\mathbf{f}$ by $\mathbf{Mf}$ and $\mathbf{\Lambda}$ by $\mathbf{\Lambda M'}$, where $\mathbf{M}$ is any orthogonal matrix of order k. In the terminology of factor analysis this corresponds to a rotation of the factors. Methods of deciding which of these choices to adopt will be fully discussed in Chapter 6. It may happen that certain elements of $\mathbf{\Lambda}$ are, by hypothesis, zero. If the pattern of zeros in $\mathbf{\Lambda}$ is such that it would be destroyed by any rotation, then $\mathbf{\Lambda}$ is uniquely determined. Here we shall suppose that there is no hypothesis of this kind. We are thus obliged to impose some more arbitrary restrictions upon the elements of $\mathbf{\Lambda}$.

Suppose that each variate is re-scaled in such a way that its residual variance, i.e. that part of the variance not due to the common factors, is unity. Then $\mathbf{\Sigma}$ becomes $\mathbf{\Psi}^{-1/2}\mathbf{\Sigma}\mathbf{\Psi}^{-1/2} = \mathbf{\Sigma}^*$ and $\mathbf{\Sigma}-\mathbf{\Psi}$ becomes

$$\mathbf{\Psi}^{-1/2}\,(\mathbf{\Sigma}-\mathbf{\Psi})\mathbf{\Psi}^{-1/2} = \mathbf{\Sigma}^*-\mathbf{I}.$$

The matrix $\mathbf{\Sigma}^*-\mathbf{I}$ is symmetric and of rank k. It may therefore be expressed in the form $\mathbf{\Omega\Delta\Omega}'$, where $\mathbf{\Delta}$ is a diagonal matrix of order

k and where $\boldsymbol{\Omega}$ is a $p \times k$ matrix satisfying $\boldsymbol{\Omega}'\boldsymbol{\Omega} = \mathbf{I}$. The elements of $\boldsymbol{\Delta}$ are the k non-zero latent roots (or eigenvalues) of $\boldsymbol{\Sigma}^* - \mathbf{I}$, while the columns of $\boldsymbol{\Omega}$ are the corresponding latent vectors in standardised form. Let us assume that the k roots are all positive and distinct and that they are arranged in decreasing order of magnitude. Then $\boldsymbol{\Omega}$ is uniquely determined (apart from possible sign reversals of its columns). We may then define $\boldsymbol{\Lambda}$ uniquely by

$$\boldsymbol{\Lambda} = \boldsymbol{\Psi}^{1/2}\boldsymbol{\Omega}\boldsymbol{\Delta}^{1/2},$$

and we have, as required,

$$\begin{aligned}\boldsymbol{\Lambda}\boldsymbol{\Lambda}' &= \boldsymbol{\Psi}^{1/2}\boldsymbol{\Omega}\boldsymbol{\Delta}\boldsymbol{\Omega}'\boldsymbol{\Psi}^{1/2} \\ &= \boldsymbol{\Psi}^{1/2}(\boldsymbol{\Sigma}^* - \mathbf{I})\boldsymbol{\Psi}^{1/2} \\ &= \boldsymbol{\Sigma} - \boldsymbol{\Psi}.\end{aligned}$$

Note that $\boldsymbol{\Sigma}^*$ has the same latent vectors as $\boldsymbol{\Sigma}^* - \mathbf{I}$, and that its p latent roots are those of $\boldsymbol{\Sigma}^* - \mathbf{I}$ increased by unity.

The above results may be summarised by the statement that if there is a unique diagonal matrix $\boldsymbol{\Psi}$ with positive elements such that the k largest latent roots of $\boldsymbol{\Sigma}^* = \boldsymbol{\Psi}^{-1/2}\boldsymbol{\Sigma}\boldsymbol{\Psi}^{-1/2}$ are distinct and greater than unity and the $p-k$ remaining roots are each unity, then $\boldsymbol{\Lambda}$ may be uniquely defined in the manner described above and the factors are, in a sense, completely identified. Though this method of defining $\boldsymbol{\Lambda}$ is arbitrary, it happens to be convenient for the estimation procedure of Chapter 4. In using the method we are in fact carrying out a principal component analysis of the matrix $\boldsymbol{\Sigma}^* - \mathbf{I}$, for a description of which see Chapter 3. Note that since

$$\boldsymbol{\Psi}^{-1/2}\boldsymbol{\Lambda} = \boldsymbol{\Omega}\boldsymbol{\Delta}^{1/2},$$

we have

$$\boldsymbol{\Lambda}'\boldsymbol{\Psi}^{-1}\boldsymbol{\Lambda} = \boldsymbol{\Delta}^{1/2}\boldsymbol{\Omega}'\boldsymbol{\Omega}\boldsymbol{\Delta}^{1/2} = \boldsymbol{\Delta}.$$

Thus we have chosen $\boldsymbol{\Lambda}$ such that $\boldsymbol{\Lambda}'\boldsymbol{\Psi}^{-1}\boldsymbol{\Lambda}$ is a diagonal matrix whose positive and distinct elements are arranged in descending order of magnitude.

A method known as the centroid or simple summation method of finding factor loadings was at one time widely used. The calculations required were relatively simple. A full description of the method has been given by Thurstone (1947) and by various other authors. We shall not discuss it in detail. Strictly speaking, the method was usually used to estimate loadings from a sample covariance or correlation matrix. Here we shall suppose that it is applied to a population matrix.

Essentially the method employs a $p \times k$ matrix $\mathbf{Q}$, of rank k, each of whose elements is either 1 or -1, the sign being in general that of the corresponding loading. Suppose that $\boldsymbol{\Psi}$ is given and that $\mathbf{\Sigma}-\mathbf{\Psi}$ is of rank k. Then the matrix $\mathbf{Q}'(\mathbf{\Sigma}-\mathbf{\Psi})\mathbf{Q}$ is a symmetric matrix of order k, which may be expressed in the form $\mathbf{TT}'$ where $\mathbf{T}$ is a lower triangular matrix. Suppose also that the diagonal elements of $\mathbf{T}$ are all real and positive. Then a loading matrix $\mathbf{\Lambda}_0$ may be defined uniquely by

$$\mathbf{\Lambda}_0 = (\mathbf{\Sigma}-\mathbf{\Psi})\mathbf{Q}\mathbf{T}'^{-1}.$$

If $\mathbf{\Lambda}$ is as previously defined, then

$$\mathbf{\Lambda}_0 = \mathbf{\Lambda}\mathbf{\Lambda}'\mathbf{Q}\mathbf{T}'^{-1} = \mathbf{\Lambda}\mathbf{M}',$$

where

$$\mathbf{M} = \mathbf{T}^{-1}\mathbf{Q}'\mathbf{\Lambda}.$$

We have

$$\mathbf{MM}' = \mathbf{T}^{-1}(\mathbf{Q}'\mathbf{\Lambda}\mathbf{\Lambda}'\mathbf{Q})\mathbf{T}'^{-1} = \mathbf{T}^{-1}(\mathbf{TT}')\mathbf{T}'^{-1} = \mathbf{I}.$$

Hence $\mathbf{M}$ is an orthogonal matrix and $\mathbf{M}'\mathbf{M} = \mathbf{I}$. Thus $\mathbf{\Lambda}_0$ may be obtained from $\mathbf{\Lambda}$, and vice versa, by rotation, and

$$\mathbf{\Lambda}_0\mathbf{\Lambda}_0' = \mathbf{\Lambda}\mathbf{M}'\mathbf{M}\mathbf{\Lambda}' = \mathbf{\Lambda}\mathbf{\Lambda}' = \mathbf{\Sigma}-\mathbf{\Psi}.$$

Suppose now that $\mathbf{\Lambda}_0$ represents *any* choice of loading matrix satisfying

$$\mathbf{\Lambda}_0\mathbf{\Lambda}_0' = \mathbf{\Sigma}-\mathbf{\Psi}.$$

Then we can without difficulty find the orthogonal matrix $\mathbf{M}$ which rotates $\mathbf{\Lambda}_0$ into $\mathbf{\Lambda}$, as previously defined, i.e. which is such that $\mathbf{\Lambda}_0\mathbf{M} = \mathbf{\Lambda}$. For we have $\mathbf{\Lambda}_0 = \mathbf{\Lambda}\mathbf{M}'$, and hence

$$\mathbf{\Lambda}_0'\mathbf{\Psi}^{-1}\mathbf{\Lambda}_0 = \mathbf{M}(\mathbf{\Lambda}'\mathbf{\Psi}^{-1}\mathbf{\Lambda})\mathbf{M}'.$$

Thus the columns of $\mathbf{M}$ are the standardised latent vectors of the matrix on the left-hand side of this equation, while the elements of the diagonal matrix $\mathbf{\Lambda}'\mathbf{\Psi}^{-1}\mathbf{\Lambda}$ are the corresponding latent roots.

2.3 UNIQUENESS OF THE PARAMETERS

Let H_k be the hypothesis that, for a specified value of k, a given covariance matrix $\mathbf{\Sigma}$ can be expressed in the form given by equation (2.2). We know that, if a suitable $\mathbf{\Psi}$ can be found, $\mathbf{\Lambda}$ can be defined uniquely in terms of $\mathbf{\Sigma}$ and $\mathbf{\Psi}$. We now consider whether a suitable $\mathbf{\Psi}$ exists, and if so, whether it is unique.

The condition, for $k > 1$, that $\mathbf{\Lambda}'\mathbf{\Psi}^{-1}\mathbf{\Lambda}$ should be diagonal has the effect of imposing $\frac{1}{2}k(k-1)$ constraints upon the parameters. Hence the number of 'free' parameters in $\mathbf{\Psi}$ and $\mathbf{\Lambda}$ is

$$p+pk-\tfrac{1}{2}k(k-1).$$

If we equate corresponding elements of the matrices on both sides of equation (2.2) we obtain $\frac{1}{2}p(p+1)$ distinct equations. Let s be the result of subtracting from this number the number of free parameters. Then

$$\begin{aligned} s &= \tfrac{1}{2}p(p+1)-p-pk+\tfrac{1}{2}k(k-1) \\ &= \tfrac{1}{2}[(p-k)^2-(p+k)]. \end{aligned}$$

If $s = 0$, we have as many equations as free parameters, so that in general we should expect H_k to be trivially true and $\mathbf{\Psi}$ to be uniquely determined. If $s < 0$, there are fewer equations than free parameters, so we should expect H_k to be trivially true, but with an infinity of choices for $\mathbf{\Psi}$ and $\mathbf{\Lambda}$. On the other hand, if $s > 0$, there are more equations than free parameters. In this case H_k is not trivial. It is true only if there are constraints upon the elements of $\mathbf{\Sigma}$. We give below some simple examples to show that in the case where $s = 0$ the above expectations are usually, but not always, fulfilled. In each example we have $p = 3$ and $k = 1$.

(a) Suppose that $\mathbf{\Sigma}$ is as follows:

1·00	0·56	0·40
	1·00	0·35
		1·00

For this, and for other symmetric matrices, we omit elements below the diagonal. To find the elements of $\mathbf{\Lambda}$, denoted here by λ_1, λ_2 and λ_3, we need only solve the equations

$$\lambda_1\lambda_2 = 0{\cdot}56, \qquad \lambda_1\lambda_3 = 0{\cdot}40, \qquad \lambda_2\lambda_3 = 0{\cdot}35.$$

Hence the elements of $\mathbf{\Lambda}$ are respectively 0·8, 0·7 and 0·5, while those of $\mathbf{\Psi}$ are 0·36, 0·51 and 0·75.There is thus a unique solution.

(b) Consider now $\mathbf{\Sigma}$ given by:

1·00	0·84	0·60
	1·00	0·35
		1·00

If we solve as before, we get

$$\lambda_1 = 1{\cdot}2, \qquad \lambda_2 = 0{\cdot}7, \qquad \lambda_3 = 0{\cdot}5,$$

and hence

$$\psi_1 = 1-\lambda_1^2 = -0{\cdot}44.$$

In this case there is no acceptable solution with $k = 1$, since there is no $\mathbf{\Psi}$ with positive elements such that the matrix $\mathbf{\Sigma}^*-\mathbf{I}$ has one positive and two zero latent roots. If we allow two factors, there is an infinity of solutions.

(c) As a further example in which there is no acceptable solution, suppose that all three non-diagonal elements of $\mathbf{\Sigma}$ are negative. Solving as before for λ_1, λ_2 and λ_3, we obtain imaginary numbers. The trouble here is that, although $\mathbf{\Psi}$ has positive elements, the only non-zero latent root of $\mathbf{\Sigma}^* - \mathbf{I}$ is negative.

(d) Finally, suppose that $\mathbf{\Sigma}$ has unit diagonal elements and zero non-diagonal elements, apart from σ_{12}, for which $0 < \sigma_{12}^2 < 1$. This allows an infinity of solutions, since λ_1 and λ_2 can have any values satisfying

$$\lambda_1\lambda_2 = \sigma_{12}, \qquad \lambda_1^2 < 1, \qquad \lambda_2^2 < 1.$$

This example violates a simple necessary condition for the uniqueness of $\mathbf{\Psi}$, that holds for any value of k, namely that every column of $\mathbf{\Lambda}$ must have at least three non-zero elements, however the factors are rotated.

The question of the uniqueness of $\mathbf{\Psi}$ has been investigated in some detail by Anderson and Rubin (1956). They have given conditions for uniqueness that are either necessary or sufficient. No condition that is both necessary and sufficient has yet been found.

2.4 VARIATES WITH NO RESIDUAL COMPONENT

The model represented by equations (2.1) and (2.2) can be generalised to some extent by allowing m diagonal elements of $\mathbf{\Psi}$, say $\psi_1, \ldots, \psi_m$, to be zero. The value of m must not exceed k, as otherwise the matrix $\mathbf{\Sigma}$ would be singular. In this case the variates $x_1, \ldots, x_m$ have no residual components and are linear functions of the common factors only. In geometrical language, these m variates lie entirely within the factor space. Hence we may choose the factors in such a way that $x_1, \ldots, x_m$ are linear functions of $f_1, \ldots, f_m$ alone. Any remaining factors then affect only the variates $x_{m+1}, \ldots, x_p$.

Let us partition the variate vector $\mathbf{x}$ into two parts, $\mathbf{x}_1$, with elements $x_1, \ldots, x_m$ and $\mathbf{x}_2$, with elements $x_{m+1}, \ldots, x_p$. The corresponding partition of $\mathbf{\Sigma}$ is

$$\begin{bmatrix} \mathbf{\Sigma}_{11} & \mathbf{\Sigma}_{12} \\ \mathbf{\Sigma}_{21} & \mathbf{\Sigma}_{22} \end{bmatrix},$$

where

$$\mathbf{\Sigma}_{11} = E(\mathbf{x}_1\mathbf{x}_1'),$$
$$\mathbf{\Sigma}_{22} = E(\mathbf{x}_2\mathbf{x}_2'),$$
$$\mathbf{\Sigma}_{12} = \mathbf{\Sigma}_{21}' = E(\mathbf{x}_1\mathbf{x}_2').$$

A similar partition of $\mathbf{\Psi}$ gives

$$\begin{bmatrix} \mathbf{0} & \mathbf{0} \\ \mathbf{0} & \mathbf{\Psi}_2 \end{bmatrix},$$

where $\mathbf{\Psi}_2$ is diagonal with elements $\psi_{m+1}, \ldots, \psi_p$.

Assume, for the present, that $m < k$. Then the factor vector $\mathbf{f}$ may be partitioned into two parts, $\mathbf{f}_1$, with elements $f_1, \ldots, f_m$ and $\mathbf{f}_2$, with elements $f_{m+1}, \ldots, f_k$. The corresponding partition of $\mathbf{\Lambda}$ is

$$\begin{bmatrix} \mathbf{\Lambda}_{11} & \mathbf{0} \\ \mathbf{\Lambda}_{21} & \mathbf{\Lambda}_{22} \end{bmatrix},$$

where the various submatrices represent the respective loading matrices of $\mathbf{x}_1$ and $\mathbf{x}_2$ on $\mathbf{f}_1$ and $\mathbf{f}_2$. Our choice of factors has made $\mathbf{\Lambda}_{12} = \mathbf{0}$. We thus have the equations

$$\mathbf{\Sigma}_{11} = \mathbf{\Lambda}_{11}\mathbf{\Lambda}'_{11},$$

$$\mathbf{\Sigma}_{21} = \mathbf{\Lambda}_{21}\mathbf{\Lambda}'_{11},$$

and

$$\mathbf{\Sigma}_{22} = \mathbf{\Lambda}_{21}\mathbf{\Lambda}'_{21}+\mathbf{\Lambda}_{22}\mathbf{\Lambda}'_{22}+\mathbf{\Psi}_2.$$

If $m = 1$, $\mathbf{\Lambda}_{11}$ is merely the scalar $\sqrt{\sigma_{11}}$. Otherwise, for given $\mathbf{\Sigma}$, the matrix $\mathbf{\Lambda}_{11}$ is arbitrary. Suppose, however, that $\mathbf{\Sigma}_{11}$ is expressed in the form $\mathbf{\Gamma}_1\mathbf{\Delta}_1\mathbf{\Gamma}'_1$, where $\mathbf{\Gamma}_1$ is an orthogonal matrix of order m and where $\mathbf{\Delta}_1$ is a diagonal matrix of order m with positive elements along the diagonal, which we assume to be distinct and arranged in decreasing order of magnitude. Then $\mathbf{\Lambda}_{11}$ may be defined uniquely by the equation

$$\mathbf{\Lambda}_{11} = \mathbf{\Gamma}_1\mathbf{\Delta}_1^{1/2},$$

and $\mathbf{\Lambda}_{21}$ is given by

$$\mathbf{\Lambda}_{21} = \mathbf{\Sigma}_{21}\mathbf{\Lambda}_{11}^{\prime -1} = \mathbf{\Sigma}_{21}\mathbf{\Gamma}_1\mathbf{\Delta}_1^{-1/2}.$$

The above choice of $\mathbf{\Lambda}_{11}$ means that the factors $\mathbf{f}_1$ are the m principal components of the variates $\mathbf{x}_1$, as found by the method given in Chapter 3.

Consider the residual covariance matrix of $\mathbf{x}_2$ when the variates $\mathbf{x}_1$, or the factors $\mathbf{f}_1$, have been eliminated. This matrix is given by

$$\mathbf{\Sigma}_{22.1} = \mathbf{\Sigma}_{22} - \mathbf{\Sigma}_{21}\mathbf{\Sigma}_{11}^{-1}\mathbf{\Sigma}_{12}.$$

Since

$$\mathbf{\Sigma}_{21}\mathbf{\Sigma}_{11}^{-1}\mathbf{\Sigma}_{12} = (\mathbf{\Lambda}_{21}\mathbf{\Lambda}'_{11})(\mathbf{\Lambda}_{11}\mathbf{\Lambda}'_{11})^{-1}(\mathbf{\Lambda}_{11}\mathbf{\Lambda}'_{21}) = \mathbf{\Lambda}_{21}\mathbf{\Lambda}'_{21},$$

we have, alternatively,

$$\mathbf{\Sigma}_{22\cdot1} = \mathbf{\Lambda}_{22}\mathbf{\Lambda}'_{22} + \mathbf{\Psi}_2 .$$

Evidently $\mathbf{\Lambda}_{22}$ and $\mathbf{\Psi}_2$ may be obtained from an analysis of the matrix $\mathbf{\Sigma}_{22\cdot1}$, in which all correlation arises from the $k-m$ factors in $\mathbf{f}_2$.

In the case where $m = k$, the factors $\mathbf{f}_2$ and the matrix $\mathbf{\Lambda}_{22}$ do not exist, and $\mathbf{\Sigma}_{22\cdot1}$ reduces to the diagonal matrix $\mathbf{\Psi}_2$.

2.5 CORRELATED FACTORS

Assume that $k > 1$. Let us now suppose that the factors in equation (2.1) may be correlated and that their variances are no longer necessarily equal to unity. The other assumptions about **f** and **e** are as before. Let ϕ_{rr} denote the variance of f_r and ϕ_{rs} the covariance of f_r and f_s. Then the covariance matrix of **f** is given by

$$E(\mathbf{ff}') = \mathbf{\Phi} = [\phi_{rs}],$$

a symmetric matrix of order k. In place of equation (2.2) we have

$$\mathbf{\Sigma} = \mathbf{\Lambda\Phi\Lambda}' + \mathbf{\Psi}. \qquad (2.3)$$

Equations (2.1) and (2.3) are still satisfied if we replace **f** by **Mf**, $\mathbf{\Lambda}$ by $\mathbf{\Lambda M}^{-1}$ and $\mathbf{\Phi}$ by $\mathbf{M\Phi M}'$, where **M** is now any non-singular matrix of order k. The matrix **M** corresponds to a non-singular transformation of the factors. Since **M** has k^2 elements, it is clear that at least k^2 constraints upon the elements of $\mathbf{\Phi}$ and $\mathbf{\Lambda}$ are necessary for uniqueness.

It is usual to suppose that the factors have unit variances, in which case k constraints are obtained by equating to unity the diagonal elements of $\mathbf{\Phi}$. The commonest method of imposing the remaining constraints is to require that certain elements of $\mathbf{\Lambda}$, at least $k-1$ in each column, are zero. If the pattern of zeros in $\mathbf{\Lambda}$ is such that it would be destroyed by any non-singular transformation of the factors, then $\mathbf{\Lambda}$ and $\mathbf{\Phi}$ are uniquely determined, given $\mathbf{\Sigma}$ and $\mathbf{\Psi}$. The idea of postulating such a pattern of zeros originated with Thurstone (1947) and was termed by him 'simple structure'. Thurstone did not specify in advance precisely which loadings were to be zero. Instead he gave rules as to their number and disposition. These rules, however, leave an element of indeterminacy that is inconvenient when estimation procedures are being developed. Hence in Chapters 6 and 7 we shall consider the formulation of hypotheses that specify the exact positions of the zeros in $\mathbf{\Lambda}$.

Exercises

2.1. Using the notation of section 2.2, suppose that there are two common factors and let $\mathbf{\Lambda}_0$ be a given matrix of loadings. Let $\mathbf{\Lambda} = \mathbf{\Lambda}_0\mathbf{M}$ be an orthogonally rotated loading matrix such that $\mathbf{\Lambda}'\mathbf{\Psi}^{-1}\mathbf{\Lambda}$ is diagonal. Expressing $\mathbf{M}$ in the form

$$\begin{bmatrix} \cos\theta & -\sin\theta \\ \sin\theta & \cos\theta \end{bmatrix},$$

show that

$$\tan 2\theta = 2\alpha_{12}/(\alpha_{11}-\alpha_{22}),$$

where α_{rs} is the element in the rth row and sth column of the matrix $\mathbf{\Lambda}_0'\mathbf{\Psi}^{-1}\mathbf{\Lambda}_0$ and $\alpha_{11} > \alpha_{22}$.

2.2. Ignoring sampling errors, take $\mathbf{\Sigma}-\mathbf{\Psi}$ to be the matrix given in *Table 6.1* of Chapter 6 with its unit diagonal elements replaced by the communalities of *Table 6.2*. Then $\mathbf{\Sigma}-\mathbf{\Psi}$ is approximately a matrix of rank 2. Taking $\mathbf{Q}$ as the matrix given by

$$\mathbf{Q}' = \begin{bmatrix} 1 & 1 & 1 & 1 & 1 & 1 \\ 1 & 1 & 1 & -1 & -1 & -1 \end{bmatrix},$$

apply the centroid method described in section 2.2 to find the 6×2 matrix of loadings $\mathbf{\Lambda}_0$.

2.3. Apply the rotation method of Exercise 2.1 to the $\mathbf{\Lambda}_0$ found in Exercise 2.2. Verify that the elements of the resulting $\mathbf{\Lambda}$ resemble closely the loadings of *Table 6.2*.

Chapter 3

PRINCIPAL COMPONENT ANALYSIS

3.1 INTRODUCTION

The main differences between factor analysis and principal component analysis have already been noted in Chapter 1. Since the latter method is relatively well known, a brief account only will be given here. In it a linear transformation is applied to the p observed variates $x_1, \ldots, x_p$ in order to produce a new set of uncorrelated and standardised variates $z_1, \ldots, z_p$. No hypothesis need be made regarding the x-variates except that they possess means and a covariance matrix $\mathbf{\Sigma}$. As Burt (1949) has pointed out, this method was in effect put forward by Karl Pearson in 1901, but its general use in practice is due to Hotelling (1933).

3.2 PRINCIPAL COMPONENTS IN THE POPULATION

Since we are interested only in the variances and covariances of variates we shall, as before, assume that the means of $x_1, \ldots, x_p$ are all zero. The p variates form a vector $\mathbf{x}$.

Let us suppose that the latent roots of the covariance matrix $\mathbf{\Sigma}$ are distinct and that they are arranged in decreasing order of magnitude. They will be denoted by $\delta_1, \ldots, \delta_p$. We may then write

$$\mathbf{\Sigma} = \mathbf{\Gamma}\mathbf{\Delta}\mathbf{\Gamma}', \tag{3.1}$$

where $\mathbf{\Delta}$ is the diagonal matrix whose diagonal elements are $\delta_1, \ldots, \delta_p$ and where $\mathbf{\Gamma}$ is an orthogonal matrix of order p. Since the latent roots of $\mathbf{\Sigma}$ are distinct, $\mathbf{\Gamma}$ is uniquely defined apart from possible sign reversals of its columns. The jth column of $\mathbf{\Gamma}$ is the latent vector corresponding to the root δ_j.

We now define new variates $y_1, \ldots, y_p$, forming a vector $\mathbf{y}$, by the equation

$$\mathbf{y} = \boldsymbol{\Gamma}'\mathbf{x}. \tag{3.2}$$

The covariance matrix of $\mathbf{y}$ is then given by

$$\begin{aligned} E(\mathbf{yy}') &= E(\boldsymbol{\Gamma}'\mathbf{xx}'\boldsymbol{\Gamma}) \\ &= \boldsymbol{\Gamma}'\boldsymbol{\Sigma}\boldsymbol{\Gamma} \\ &= \boldsymbol{\Gamma}'(\boldsymbol{\Gamma}\boldsymbol{\Delta}\boldsymbol{\Gamma}')\boldsymbol{\Gamma} \\ &= \boldsymbol{\Delta} \qquad (\text{since } \boldsymbol{\Gamma}'\boldsymbol{\Gamma} = \mathbf{I}). \end{aligned}$$

Thus the new variates are uncorrelated and the variance of y_j is δ_j. It is easy to show that y_1 is the normalised linear combination of $x_1, \ldots, x_p$ with maximum variance. For any such combination may be written as $\mathbf{a}'\mathbf{x}$, where $\mathbf{a}$ is a column vector satisfying $\mathbf{a}'\mathbf{a} = 1$. It may also be written as $\mathbf{c}'\mathbf{y} = \mathbf{c}'\boldsymbol{\Gamma}'\mathbf{x}$, where

$$\mathbf{a} = \boldsymbol{\Gamma}\mathbf{c}, \qquad \mathbf{c} = \boldsymbol{\Gamma}'\mathbf{a},$$

and where

$$\mathbf{c}'\mathbf{c} = \mathbf{a}'\boldsymbol{\Gamma}\boldsymbol{\Gamma}'\mathbf{a} = \mathbf{a}'\mathbf{a} = 1.$$

Hence we may equally well consider normalised linear combinations of $y_1, \ldots, y_p$. Now the variance of $\mathbf{c}'\mathbf{y} = \sum_j c_j y_j$ is

$$\sum_{j=1}^{p} c_j^2 \delta_j = \delta_1 + \sum_{j=2}^{p} c_j^2(\delta_j - \delta_1),$$

since $\sum c_j^2 = 1$. This variance is clearly maximised when we take $c_2 = \ldots = c_p = 0$ and $c_1 = 1$. Similarly y_2 is the normalised combination of the x-variates that has maximum variance subject to being uncorrelated with y_1. For such a combination must be of the form $c_2y_2 + \ldots + c_py_p$ and for maximum variance we must have $c_2 = 1$ and the remaining coefficients zero. Similarly for the remaining y variates. The variance of each y_j is maximised subject to its being uncorrelated with all predecessors.

The y-variates are thus orthogonal both in the statistical sense of being uncorrelated and in the mathematical sense that the coefficients of each linear combination form a column of the orthogonal matrix $\boldsymbol{\Gamma}$. The trace, or sum of diagonal elements, of $\boldsymbol{\Sigma}$ is the same as that of $\boldsymbol{\Delta}$, that is

$$\text{tr}\,(\boldsymbol{\Sigma}) = \text{tr}\,(\boldsymbol{\Delta}) = \sum_j \delta_j.$$

This means that the total variance of the y-variates is the same as that of the original x-variates.

Principal components are often taken to be standardised variates, having unit variances. We therefore define the jth component, for

$j = 1, \ldots, p$, as $z_j = y_j/\sqrt{\delta_j}$. In matrix notation this becomes

$$\mathbf{z} = \mathbf{\Delta}^{-1/2}\mathbf{y}, \tag{3.3}$$

where $\mathbf{z}$ is the column vector with elements $z_1, \ldots, z_p$. On combining (3.2) and (3.3) we have

$$\mathbf{z} = \mathbf{\Delta}^{-1/2}\mathbf{\Gamma}'\mathbf{x}. \tag{3.4}$$

The inverse relationship giving $\mathbf{x}$ in terms of $\mathbf{z}$ is

$$\mathbf{x} = \mathbf{\Gamma}\mathbf{\Delta}^{1/2}\mathbf{z}. \tag{3.5}$$

An important difference between this equation and the basic equation (2.1) of factor analysis is that in (3.5) there is no residual vector $\mathbf{e}$. As we have seen, in principal component analysis the total variance of the variates is accounted for only when all p components have been found.

Instead of assuming that all p latent roots of $\mathbf{\Sigma}$ are distinct we may suppose that only the first k roots are distinct, where $k < p-1$, and that the remaining $p-k$ roots are equal to some value δ. Then only the first k columns of $\mathbf{\Gamma}$ are uniquely defined, since there is an infinity of choices for the remaining $p-k$ columns such that $\mathbf{\Gamma}$ is an orthogonal matrix. In this case only the first k principal components need be found, as all normalised linear combinations of the x-variates orthogonal to these have the same variance δ.

3.3 SAMPLE ESTIMATES

We now consider principal component analysis as applied to a sample. We suppose that a random sample of $N(> p)$ sets of observations of the p variates in $\mathbf{x}$ has been obtained. From this the usual unbiased sample estimates of the variances and covariances are found. These form a symmetric and positive definite matrix $\mathbf{S}$ defined by

$$\mathbf{S} = (1/n)\left(\sum_{\alpha} \mathbf{x}_\alpha \mathbf{x}'_\alpha - N\bar{\mathbf{x}}\bar{\mathbf{x}}'\right),$$

where $\mathbf{x}_\alpha(\alpha = 1, \ldots, N)$ is the column vector representing the α-th set of observations, $\bar{\mathbf{x}}$ is the sample mean vector $(1/N)\sum \mathbf{x}_\alpha$ and $n = N-1$. Principal components for the sample are defined in exactly the same way as for the population, using $\mathbf{S}$ in place of $\mathbf{\Sigma}$, and they have analogous properties. The sample equivalent of equation (3.1) is

$$\mathbf{S} = \mathbf{G}\mathbf{D}\mathbf{G}', \tag{3.6}$$

where **D** is the diagonal matrix whose diagonal elements $d_1, \ldots, d_p$ are the latent roots of **S**, supposed distinct and arranged in decreasing order of magnitude, and where **G** is an orthogonal matrix. The jth column of **G** is the latent vector corresponding to the root d_j. For the sample principal components we substitute **G** for $\mathbf{\Gamma}$ and **D** for $\mathbf{\Delta}$ in equations (3.2)–(3.5).

If n is sufficiently large, the sample root d_j, for $j = 1, \ldots, p$, corresponds almost certainly to the population root δ_j. For each value of j we may suppose that a specified element of the jth column of $\mathbf{\Gamma}$ is taken to be positive and that so also is the corresponding element of the jth column of **G**. With some such rule for determining the signs of their elements, there is then a complete correspondence between **G** and $\mathbf{\Gamma}$.

We shall assume in future that **x** has a multivariate normal distribution. The distribution of the elements of **S** under this assumption is discussed in section 4.2 of Chapter 4. If we consider only the information provided by **S** (and not that provided by $\bar{\mathbf{x}}$), the elements of **S** are maximum likelihood estimates of the elements of $\mathbf{\Sigma}$. By standard estimation theory it follows that **D** and **G** provide maximum likelihood estimates of $\mathbf{\Delta}$ and $\mathbf{\Gamma}$ respectively.

3.4 PRACTICAL APPLICATIONS

Principal component analysis is most useful when the variates are all measured in the same units. If they are not, the method is more difficult to justify. A change in the scales of measurement of some or all of the variates results in the covariance matrix being multiplied on both sides by a diagonal matrix. The effect of this on the latent roots and vectors is complicated and the components are not invariant under such changes of scale. In this respect principal component analysis compares unfavourably with the maximum likelihood factor analysis methods discussed in other chapters, which are invariant.

In educational and psychological work it is a common practice to standardise each variate. The sample covariance matrix **S** is then replaced by the correlation matrix **R** having unities in the diagonal and correlation coefficients elsewhere. This causes difficulties if the distributional properties of estimates are of interest or if significance tests are required. Though a correlation matrix is used in the numerical example of the next section, this should not be taken as an endorsement of the procedure.

It is common practice, especially if p is large, to make use of only the first few latent roots and vectors. If these account for a high proportion of the total variance, the remaining ones are of no practical importance. In some such cases the component analysis may be a prelude to some other form of multivariate analysis in which the original variates are replaced by a much smaller number of components.

3.5 A NUMERICAL EXAMPLE

To illustrate the application of principal component analysis to observational data we use the correlation coefficients between five psychophysical measurements for a sample of 123 individuals. The correlation matrix **R**, with sub-diagonal elements omitted, is given in *Table 3.1*.

Table 3.1 CORRELATION MATRIX FOR FIVE PSYCHOPHYSICAL MEASUREMENTS

1	*2*	*3*	*4*	*5*
1·000	0·438	−0·137	0·205	−0·178
	1·000	0·031	0·180	−0·304
		1·000	0·161	0·372
			1·000	−0·013
				1·000

The latent roots and vectors are as shown in *Table 3.2*. The latter are given as row vectors and thus form the matrix **G**′. The largest element in each has been made positive. We shall not discuss here the method by which the roots and vectors were computed. Some brief remarks on this topic will be found in Appendix I.

Table 3.2 LATENT ROOTS AND VECTORS

Roots	*Row vectors*				
1·7571	0·5555	0·5647	−0·2700	0·2357	−0·4940
1·3307	0·1857	0·2475	0·6620	0·5565	0·3954
0·7809	−0·2160	−0·4397	−0·3204	0·7870	−0·1948
0·7092	0·6408	−0·3076	−0·3984	−0·0148	0·5795
0·4221	−0·4469	0·5761	−0·4770	0·1227	0·4752

The variates are given in terms of the components by the equation $\mathbf{x} = \mathbf{Wz}$, where $\mathbf{W} = \mathbf{GD}^{1/2}$. To find the matrix **W** of 'weights' we multiply each row vector in *Table 3.2* by the square root of the corresponding latent root and transpose it, so that rows become columns. The matrix **W** is given in *Table 3.3*. The calculations may be checked by verifying that $\mathbf{WW'} = \mathbf{R}$ and that $\mathbf{W'W} = \mathbf{D}$.

Table 3.3 MATRIX OF WEIGHTS

	Components				
Variates	I	II	III	IV	V
1	0·736	0·214	−0·191	0·540	−0·290
2	0·749	0·285	−0·389	−0·259	0·374
3	−0·358	0·764	−0·283	−0·335	−0·310
4	0·312	0·642	0·695	−0·012	0·080
5	−0·655	0·456	−0·172	0·488	0·309

In terms of the components x_1 is, for example, given by

$$x_1 = 0{\cdot}736z_1 + 0{\cdot}214z_2 - 0{\cdot}191z_3 + 0{\cdot}540z_4 - 0{\cdot}290z_5.$$

If we require to express the components in terms of the variates, we make use of the columns of *Table 3.3*, the figures in each column being divided by the corresponding latent root. Thus, for example, we have

$$\begin{aligned} z_1 &= (0{\cdot}736x_1 + \ldots - 0{\cdot}655x_5)/1{\cdot}7571 \\ &= 0{\cdot}419x_1 + 0{\cdot}426x_2 - 0{\cdot}204x_3 + 0{\cdot}178x_4 - 0{\cdot}373x_5. \end{aligned}$$

By expressing the latent roots as percentages of the total variance in **x**, in this case 5, we find that the components account successively for 35·1, 26·6, 15·6, 14·2 and 8·4 per cent of the total variance.

3.6 TESTING EQUALITY OF LATENT ROOTS

Though it is mainly the sizes of the latent roots that are of interest, we may sometimes wish to test the hypothesis that the $p-k$ smallest latent roots of $\mathbf{\Sigma}$ are equal. If this hypothesis is accepted, there is no point in finding more than k components from the data. We shall henceforth assume that the sample size is reasonably large. A safe rule is to require that $n-p > 50$.

We consider first the simple case where $k = 0$. The hypothesis that all p latent roots of $\mathbf{\Sigma}$ are equal is the same as the hypothesis that the x-variates are independent of one another and that they have the same (unknown) variance. We suppose that an unstandardised sample covariance matrix **S** has been analysed. In this case, as Bartlett (1954) has shown, an appropriate criterion for testing the hypothesis is

$$[n-(1/6)(2p+1+2/p)][-\log_e|\mathbf{S}|+p\log_e(\operatorname{tr}\mathbf{S}/p)]. \qquad (3.7)$$

For moderately large values of n this is distributed under the hypothesis very nearly as χ^2 with $\frac{1}{2}(p+2)(p-1)$ degrees of freedom.

Now suppose that $0 < k < p-1$. For testing the hypothesis that the $p-k$ smallest latent roots of $\mathbf{\Sigma}$ are equal Bartlett suggested as an approximate χ^2 criterion the statistic

$$n'[-\log|\mathbf{S}|+\log(d_1d_2\ldots d_k)+q\log d], \qquad (3.8)$$

where $q = p-k$ and where

$$d = (\operatorname{tr}\mathbf{S}-d_1-\ldots-d_k)/q.$$

With sufficient accuracy the value of n' is given by

$$n' = n-k-(1/6)(2q+1+2/q),$$

though some results of Lawley (1956) suggest that the χ^2 approximation is slightly improved if the above value is increased by an amount

$$d^2\sum_{j=1}^{k} 1/(d_j-d)^2.$$

The number of degrees of freedom for χ^2 in this case is equal to $\frac{1}{2}(q+2)(q-1)$.

We next consider the problems that arise when the variates have all been standardised and when $\mathbf{S}$ is replaced by the sample correlation matrix $\mathbf{R}$. In the case where $k = 0$ the hypothesis to be tested is that all p latent roots of the population correlation matrix $\mathbf{P}$ are equal to unity. This can happen only if $\mathbf{P} = \mathbf{I}$. Hence the hypothesis is that the variates are independent of one another. A suitable test criterion was shown by Box (1949) to be

$$-[n-(2p+5)/6]\log_e|\mathbf{R}|. \qquad (3.9)$$

Under the hypothesis of independence this is distributed very nearly as χ^2 with $\frac{1}{2}p(p-1)$ degrees of freedom.

Considerable difficulties arise when a test is required of the hypothesis that, for $0 < k < p-1$, the $p-k$ smallest latent roots of $\mathbf{P}$ are equal. The test criterion that Bartlett (1951) suggested is

$$n[-\log|\mathbf{R}|+\log(d_1d_2\ldots d_k)+(p-k)\log d], \qquad (3.10)$$

where d_j now denotes the jth latent root of $\mathbf{R}$ and where

$$d = (p-d_1-\ldots-d_k)/(p-k).$$

Unfortunately when $n \to \infty$ the distribution of this criterion under the hypothesis does not tend to the χ^2 form. However, if the first k components account for a fairly high proportion of the total variance or if we are prepared to accept a very rough test, we may treat expression (3.10) as a χ^2 variate with $\frac{1}{2}(p-k+2)(p-k-1)$ degrees of freedom. The multiplying factor for the expression in square brackets has been taken simply as n since nothing more sophisticated is justifiable.

To illustrate this test we consider the data of section 3.5 and test the hypothesis that the three smallest latent roots of $\mathbf{P}$ are equal. We have $n = 122$, $p = 5$, $k = 2$ and

$$|\mathbf{R}| = d_1 d_2 \ldots d_5.$$

The criterion (3.10) is evaluated as

$$\begin{aligned} 122\,[-\log_e(d_3 d_4 d_5) + 3 \log_e d] \\ = 122(-\log 0{\cdot}2338 + 3 \log 0{\cdot}6374) \\ = 12{\cdot}5. \end{aligned}$$

The number of degrees of freedom is 5, so this value is considered to be significant at the 5 per cent level. We are thus inclined to reject the hypothesis and to suspect that the three smallest roots are genuinely unequal.

The case where $k = 1$ has received some attention in the literature on the subject. The hypothesis that the $p-1$ smallest latent roots of a correlation matrix of order p are equal is the same as the hypothesis that all correlation coefficients between the variates are equal. The asymptotic distribution of the criterion (3.10) for this case was found by Lawley (1963). He also proposed an alternative criterion having an asymptotic χ^2 distribution under the hypothesis. This is not, however, a function of the latent roots of $\mathbf{R}$. If the hypothesis of equal correlation is assumed to be true, the first principal component should be taken as a constant times $\sum_i x_i$.

3.7 SAMPLING FORMULAE FOR SAMPLE ESTIMATES

It is possible to find the asymptotic variances and covariances of the latent roots d_j of $\mathbf{S}$. Assuming that n is large, the variance of d_j is given by

$$\text{var}(d_j) = 2\delta_j^2/n.$$

This formula could be used to provide approximate confidence limits for δ_j. If δ_j and δ_l are two distinct roots of $\mathbf{\Sigma}$, the two corresponding roots of $\mathbf{S}$, d_j and d_l, are asymptotically uncorrelated. These results were first obtained by Girshick (1939).

It is also possible to obtain formulae for the asymptotic variances and covariances of component weights, assuming that the analysis has been carried out on the unstandardised matrix **S** and that all latent roots of $\mathbf{\Sigma}$ are distinct. These formulae may be derived by use of the methods of Chapter 5. It seems unlikely that they would be of much practical value, but we give the results for what they are worth. For clarity we now denote the matrix of weights in the population by **W** and that in the sample by $\hat{\mathbf{W}}$. Thus we have

$$\mathbf{W} = \mathbf{\Gamma\Delta}^{1/2},$$
$$\hat{\mathbf{W}} = \mathbf{GD}^{1/2}.$$

The element in the ith row and jth column of **W** is denoted by w_{ij}. This is the true weight of the component z_j in the expression for x_i. The corresponding element of $\hat{\mathbf{W}}$ is $\hat{w}_{ij}$.

For the covariance of $\hat{w}_{ij}$ and $\hat{w}_{hj}$ we have

$$n \operatorname{cov}(\hat{w}_{ij}, \hat{w}_{hj}) = \tfrac{1}{2}w_{ij}w_{hj} + \delta_j^2 \sum_l{}' [w_{il}w_{hl}/(\delta_j - \delta_l)^2],$$

where $\sum'$ denotes summation of l over all values from 1 to p except j. For the variance of $\hat{w}_{ij}$ we put $h = i$. The covariance of w_{ij} and $\hat{w}_{hl}$ for $j \neq l$ is given by

$$n \operatorname{cov}(\hat{w}_{ij}, \hat{w}_{hl}) = -\delta_j\delta_l w_{il}w_{hj}/(\delta_j - \delta_l)^2.$$

Variances and covariances of the latent roots of a sample correlation matrix **R** have been found by Girshick (1939). They are much more complicated in form than the corresponding results for **S**. This would apply also to sampling formulae for the weights derived from **R** and there seems to be little point in attempting to obtain such formulae. In some circumstances the results obtained for the unstandardised case might serve as rough approximations, for instance if the observed variates were measured on scales for which the true variances were nearly equal.

Chapter 4

ESTIMATION IN THE UNRESTRICTED MODEL

4.1 INTRODUCTION

We return to the factor model represented by equations (2.1) and (2.2), in which the factors are uncorrelated. Since only arbitrary constraints are imposed upon the parameters to define them uniquely, this will be termed the 'unrestricted' model. We assume that, for specified k, the hypothesis H_k that there are k common factors is non-trivial. We suppose that there is a unique $\mathbf{\Psi}$, with positive diagonal elements, and a unique $\mathbf{\Lambda}$ such that $\mathbf{\Lambda}'\mathbf{\Psi}^{-1}\mathbf{\Lambda}$ is a diagonal matrix whose diagonal elements are positive, distinct and arranged in decreasing order of magnitude. We consider the problem of estimating $\mathbf{\Lambda}$ and $\mathbf{\Psi}$ from a sample covariance matrix by use of the method of maximum likelihood.

This method was first applied to factor analysis by Lawley (1940, 1941, 1943). Unfortunately the iterative computational method that he then proposed for solving the equations satisfied by the estimates had numerous disadvantages. Other methods of computation were subsequently suggested by various authors, of which the most generally successful was that employed by Howe (1955) and later by Bargmann (1957). This was a Gauss–Seidel type of iterative procedure. None of the methods was entirely satisfactory, however, since even with the best of them convergence to the final solution was often rather slow, and in some cases the procedure actually diverged. In using Howe's method Browne (1965) introduced a modification that effectively prevented divergence, but the number of iterations required still tended to be rather large. In view of these difficulties, it is not suprising that applications of maximum likelihood estimation have been rare in the literature of factor analysis.

Recently an entirely new method has been developed. It is described in an expository paper by Jöreskog and Lawley (1968) and, in more technical detail, by Jöreskog (1967) and by Lawley (1967). The method has so far proved completely successful. A computer program for it is available in Fortran IV (Jöreskog, 1966a). The new method, though iterative, converges extremely rapidly, and the values of the parameters that maximise the likelihood can be determined as accurately as desired.

4.2 MAXIMUM LIKELIHOOD ESTIMATION

We begin with the assumption that the vector variables $\mathbf{f}$ and $\mathbf{e}$ follow independent multivariate normal distributions with zero mean vectors and with respective covariance matrices $\mathbf{I}_k$ and $\mathbf{\Psi}$. The properties of the multivariate normal distribution have been described in books by various authors. See, for example, that by Anderson (1958). Since $\mathbf{x}$ is expressed in terms of $\mathbf{f}$ and $\mathbf{e}$ by equation (2.1), it follows that $\mathbf{x}$ also has a multivariate normal distribution with zero mean vector. Its covariance matrix $\mathbf{\Sigma}$ is given by equation (2.2).

Suppose that a random sample of N $(>p)$ observations of $\mathbf{x}$ is obtained, denoted by column vectors $\mathbf{x}_\alpha(\alpha = 1, \ldots, N)$. Then we define the sample covariance matrix $\mathbf{S} = [s_{ij}]$ by

$$\begin{aligned}\mathbf{S} &= (1/n)\sum_\alpha (\mathbf{x}_\alpha - \bar{\mathbf{x}})(\mathbf{x}_\alpha - \bar{\mathbf{x}})' \\ &= (1/n)\Big(\sum_\alpha \mathbf{x}_\alpha \mathbf{x}_\alpha' - N\bar{\mathbf{x}}\bar{\mathbf{x}}'\Big),\end{aligned}$$

where $n = N-1$ and where $\bar{\mathbf{x}}$ is the sample mean vector $(1/N)\sum_\alpha \mathbf{x}_\alpha$. This definition of $\mathbf{S}$ has the advantage that s_{ij} is an unbiased estimator of σ_{ij}, and $E(\mathbf{S}) = \mathbf{\Sigma}$. The elements of $\mathbf{S}$, or of $n\mathbf{S}$, are said to follow a Wishart distribution with n degrees of freedom. In Anderson's (1958) notation, the distribution of $n\mathbf{S}$ is $W(\mathbf{\Sigma}, n)$.

In factor analysis it is usual to discard the sample mean vector $\bar{\mathbf{x}}$ and to make use of the matrix $\mathbf{S}$ alone. The likelihood function L based only on the information provided by $\mathbf{S}$ is, omitting a function of the observations, given by

$$\log_e L = -\tfrac{1}{2}n \log_e |\mathbf{\Sigma}| - \tfrac{1}{2}n \sum_{i,j} s_{ij}\sigma^{ij},$$

where σ^{ij} is the element in the ith row and jth column of $\mathbf{\Sigma}^{-1}$. We may alternatively write

$$\log_e L = -\tfrac{1}{2}n\,[\log_e |\mathbf{\Sigma}| + \mathrm{tr}\,(\mathbf{S}\mathbf{\Sigma}^{-1})]. \tag{4.1}$$

The likelihood is to be regarded as a function of the elements of $\mathbf{\Sigma}$ and hence as a function of the elements of $\mathbf{\Lambda}$ and of $\mathbf{\Psi}$. We now treat these not as representing true values of the parameters but as mathematical variables that are free to vary over the complete parameter space. We seek values of $\mathbf{\Lambda}$ and $\mathbf{\Psi}$, denoted eventually by $\hat{\mathbf{\Lambda}}$ and $\hat{\mathbf{\Psi}}$, that maximise the value of L and hence that of $\log_e L$.

For reasons that will become apparent later it is more convenient to minimise the function.

$$F_k(\mathbf{\Lambda}, \mathbf{\Psi}) = \log |\mathbf{\Sigma}| + \text{tr}\,(\mathbf{S}\mathbf{\Sigma}^{-1}) - \log |\mathbf{S}| - p. \qquad (4.2)$$

This is equivalent to maximising L, since $\log L$ is $-\frac{1}{2}n$ times F_k plus a function of the observations only. We have attached a suffix k to F_k in order to indicate that $\mathbf{\Lambda}$ has k columns. The suffix will often be omitted where this can be done without ambiguity.

For the purpose of minimising the function F we require its partial derivatives with respect to the elements of $\mathbf{\Lambda}$ and the diagonal elements of $\mathbf{\Psi}$. From Appendix I, equations (A1.13) and (A1.14), we have

$$\partial \log |\mathbf{\Sigma}|/\partial\sigma_{ii} = \sigma^{ii},$$
$$\partial \log |\mathbf{\Sigma}|/\partial\sigma_{ij} = 2\sigma^{ij} \qquad (i \neq j),$$
$$\partial\, \text{tr}\,(\mathbf{S}\mathbf{\Sigma}^{-1})/\partial\sigma_{ii} = -(\mathbf{\Sigma}^{-1}\mathbf{S}\mathbf{\Sigma}^{-1})_{ii},$$
$$\partial\, \text{tr}\,(\mathbf{S}\mathbf{\Sigma}^{-1})/\partial\sigma_{ij} = -2(\mathbf{\Sigma}^{-1}\mathbf{S}\mathbf{\Sigma}^{-1})_{ij} \qquad (i \neq j),$$

where $(\mathbf{\Sigma}^{-1}\mathbf{S}\mathbf{\Sigma}^{-1})_{ij}$ denotes the element in the ith row and jth column of the matrix inside the brackets. A similar notation is used elsewhere. Hence we have

$$\partial F/\partial\sigma_{ii} = (\mathbf{\Sigma}^{-1} - \mathbf{\Sigma}^{-1}\mathbf{S}\mathbf{\Sigma}^{-1})_{ii},$$
$$\partial F/\partial\sigma_{ij} = 2(\mathbf{\Sigma}^{-1} - \mathbf{\Sigma}^{-1}\mathbf{S}\mathbf{\Sigma}^{-1})_{ij} \qquad (i \neq j).$$

Since

$$\sigma_{ii} = \sum_s \lambda_{is}^2 + \psi_i,$$
$$\sigma_{ij} = \sum_s \lambda_{is}\lambda_{js} \qquad (i \neq j),$$

we also have

$$\partial\sigma_{ii}/\partial\lambda_{ir} = 2\lambda_{ir},$$
$$\partial\sigma_{ij}/\partial\lambda_{ir} = \lambda_{jr} \qquad (i \neq j),$$
$$\partial\sigma_{ii}/\partial\psi_i = 1.$$

Hence

$$\partial F/\partial\lambda_{ir} = \sum_j [(\partial F/\partial\sigma_{ij})(\partial\sigma_{ij}/\partial\lambda_{ir})]$$
$$= 2\sum_j [\mathbf{\Sigma}^{-1}(\mathbf{\Sigma} - \mathbf{S})\mathbf{\Sigma}^{-1}]_{ij}\lambda_{jr}$$
$$= 2\,[\mathbf{\Sigma}^{-1}(\mathbf{\Sigma} - \mathbf{S})\mathbf{\Sigma}^{-1}\mathbf{\Lambda}]_{ir}$$

and

$$\partial F/\partial \psi_i = (\partial F/\partial \sigma_{ii})(\partial \sigma_{ii}/\partial \psi_i)$$
$$= [\mathbf{\Sigma}^{-1}(\mathbf{\Sigma}-\mathbf{S})\mathbf{\Sigma}^{-1}]_{ii}.$$

These results may be expressed more concisely (see section A1.16 of Appendix I) in the form

$$\partial F/\partial \mathbf{\Lambda} = 2\mathbf{\Sigma}^{-1}(\mathbf{\Sigma}-\mathbf{S})\mathbf{\Sigma}^{-1}\mathbf{\Lambda} \tag{4.3}$$

and

$$\partial F/\partial \mathbf{\Psi} = \text{diag}\,[\mathbf{\Sigma}^{-1}(\mathbf{\Sigma}-\mathbf{S})\mathbf{\Sigma}^{-1}], \tag{4.4}$$

where diag (**X**) represents the diagonal matrix formed from **X** by replacing all non-diagonal elements of **X** by zeros.

In finding the minimum of F_k it is convenient to use a two-stage procedure. First we find the conditional minimum for given $\mathbf{\Psi}$, denoted by $f_k(\mathbf{\Psi})$. Then we find the $\mathbf{\Psi}$ that minimises f_k. As we shall verify, it is possible, under certain conditions, to obtain a $\mathbf{\Lambda}$ which is such that $\partial F/\partial \mathbf{\Lambda} = \mathbf{0}$ and also such that $\mathbf{\Lambda}'\mathbf{\Psi}^{-1}\mathbf{\Lambda}$ is diagonal. This $\mathbf{\Lambda}$ both minimises F for given $\mathbf{\Psi}$ and satisfies the constraint that we imposed upon the true $\mathbf{\Lambda}$ to obtain uniqueness.

Equating $\partial F/\partial \mathbf{\Lambda}$ to zero gives

$$\mathbf{\Sigma}^{-1}(\mathbf{\Sigma}-\mathbf{S})\mathbf{\Sigma}^{-1}\mathbf{\Lambda} = \mathbf{0},$$

from which, pre-multiplying by $\mathbf{\Sigma}$, we obtain

$$(\mathbf{\Sigma}-\mathbf{S})\mathbf{\Sigma}^{-1}\mathbf{\Lambda} = \mathbf{0}. \tag{4.5}$$

To simplify this further we make use of the identity

$$\mathbf{\Sigma}^{-1} = \mathbf{\Psi}^{-1}-\mathbf{\Psi}^{-1}\mathbf{\Lambda}(\mathbf{I}+\mathbf{\Delta})^{-1}\mathbf{\Lambda}'\mathbf{\Psi}^{-1}, \tag{4.6}$$

where

$$\mathbf{\Delta} = \mathbf{\Lambda}'\mathbf{\Psi}^{-1}\mathbf{\Lambda}.$$

This identity may readily be verified if we post-multiply the right-hand side by $\mathbf{\Sigma}$ in the form $\mathbf{\Lambda}\mathbf{\Lambda}'+\mathbf{\Psi}$. Post-multiplication by $\mathbf{\Lambda}$ gives the further identity

$$\mathbf{\Sigma}^{-1}\mathbf{\Lambda} = \mathbf{\Psi}^{-1}\mathbf{\Lambda}(\mathbf{I}+\mathbf{\Delta})^{-1}. \tag{4.7}$$

By use of this, equation (4.5) may be written in the form

$$(\mathbf{\Sigma}-\mathbf{S})\mathbf{\Psi}^{-1}\mathbf{\Lambda}(\mathbf{I}+\mathbf{\Delta})^{-1} = \mathbf{0}. \tag{4.8}$$

Post-multiplying by $(\mathbf{I}+\mathbf{\Delta})$ and writing $\mathbf{\Sigma}$ as $\mathbf{\Lambda}\mathbf{\Lambda}'+\mathbf{\Psi}$, we obtain

$$(\mathbf{\Lambda}\mathbf{\Lambda}'+\mathbf{\Psi}-\mathbf{S})\mathbf{\Psi}^{-1}\mathbf{\Lambda} = \mathbf{0},$$

which simplifies to

$$\mathbf{S}\mathbf{\Psi}^{-1}\mathbf{\Lambda} = \mathbf{\Lambda}(\mathbf{I}+\mathbf{\Delta}). \tag{4.9}$$

Let the $\boldsymbol{\Lambda}$ that satisfies (4.9) be denoted by $\boldsymbol{\Lambda}_0$. Then pre-multiplying by $\boldsymbol{\Psi}^{-1/2}$, we have

$$(\boldsymbol{\Psi}^{-1/2}\mathbf{S}\boldsymbol{\Psi}^{-1/2})(\boldsymbol{\Psi}^{-1/2}\boldsymbol{\Lambda}_0) = (\boldsymbol{\Psi}^{-1/2}\boldsymbol{\Lambda}_0)(\mathbf{I}+\boldsymbol{\Delta}_0), \tag{4.10}$$

where

$$\boldsymbol{\Delta}_0 = \boldsymbol{\Lambda}_0'\boldsymbol{\Psi}^{-1}\boldsymbol{\Lambda}_0.$$

Since $\boldsymbol{\Delta}_0$ is diagonal, it is clear on comparing (4.10) with equation (A1.6) of Appendix I that the columns of $\boldsymbol{\Psi}^{-1/2}\boldsymbol{\Lambda}_0$ are latent vectors of the matrix $\mathbf{S}^*$ defined by

$$\mathbf{S}^* = \boldsymbol{\Psi}^{-1/2}\mathbf{S}\boldsymbol{\Psi}^{-1/2}. \tag{4.11}$$

The diagonal elements of $\mathbf{I}+\boldsymbol{\Delta}_0$ are the corresponding latent roots. We shall show later that, as intuition would suggest, these must be the k largest roots.

Let us suppose that the k largest latent roots $\theta_1, \ldots, \theta_k$ of $\mathbf{S}^*$ are distinct and greater than unity. The probability that this will be so tends to unity as the sample size $N \to \infty$. Suppose also that $\theta_1, \ldots, \theta_k$ are arranged in decreasing order of magnitude. Let the corresponding latent column vectors, now in standardised form, be denoted by $\boldsymbol{\omega}_1, \ldots, \boldsymbol{\omega}_k$. Denote by $\boldsymbol{\Theta}$ the diagonal matrix with $\theta_1, \ldots, \theta_k$ as diagonal elements and by $\boldsymbol{\Omega}$ the $p \times k$ matrix with columns $\boldsymbol{\omega}_1, \ldots, \boldsymbol{\omega}_k$. Then $\boldsymbol{\Omega}$ is such that $\boldsymbol{\Omega}'\boldsymbol{\Omega} = \mathbf{I}_k$ and it satisfies the equation

$$\mathbf{S}^*\boldsymbol{\Omega} = \boldsymbol{\Omega}\boldsymbol{\Theta}.$$

This is equivalent to (4.10) if we put

$$\boldsymbol{\Psi}^{-1/2}\boldsymbol{\Lambda}_0 = \boldsymbol{\Omega}(\boldsymbol{\Theta}-\mathbf{I})^{1/2} \tag{4.12}$$

and

$$\boldsymbol{\Delta}_0 = \boldsymbol{\Lambda}_0'\boldsymbol{\Psi}^{-1}\boldsymbol{\Lambda}_0 = \boldsymbol{\Theta}-\mathbf{I}.$$

Thus $\boldsymbol{\Lambda}_0$ is defined by

$$\boldsymbol{\Lambda}_0 = \boldsymbol{\Psi}^{1/2}\boldsymbol{\Omega}(\boldsymbol{\Theta}-\mathbf{I})^{1/2}. \tag{4.13}$$

This definition breaks down if any two of $\theta_1, \ldots, \theta_k$ are equal, as the corresponding vectors are then indeterminate. Even near-equality causes trouble, since it makes the vectors difficult to compute. Fortunately such cases are rare in practice. It is possible for one or more of the roots to be less than unity, corresponding to imaginary elements in $\boldsymbol{\Lambda}$, but this also rarely happens. In the computer program for the above procedure any such root is replaced by unity. The corresponding column of $\boldsymbol{\Lambda}$ then becomes null.

With $\boldsymbol{\Lambda}_0$ as above we have

$$f_k(\boldsymbol{\Psi}) = \min_{\boldsymbol{\Lambda}} F_k(\boldsymbol{\Lambda}, \boldsymbol{\Psi}) = F_k(\boldsymbol{\Lambda}_0, \boldsymbol{\Psi}).$$

Hence, using (4.2), we find that

$$f_k(\boldsymbol{\Psi}) = \log|\boldsymbol{\Sigma}_0| + \operatorname{tr}(\mathbf{S}\boldsymbol{\Sigma}_0^{-1}) - \log|\mathbf{S}| - p, \tag{4.14}$$

where

$$\boldsymbol{\Sigma}_0 = \boldsymbol{\Lambda}_0\boldsymbol{\Lambda}_0' + \boldsymbol{\Psi}.$$

We now express f_k in terms of the $(p-k)$ smallest latent roots $\theta_{k+1}, \ldots, \theta_p$ of $\mathbf{S}^*$. Consider the matrix

$$\boldsymbol{\Psi}^{-1/2}\boldsymbol{\Sigma}_0\boldsymbol{\Psi}^{-1/2} = \mathbf{I} + \boldsymbol{\Psi}^{-1/2}\boldsymbol{\Lambda}_0\boldsymbol{\Lambda}_0'\boldsymbol{\Psi}^{-1/2} = \mathbf{I} + \boldsymbol{\Omega}(\boldsymbol{\Theta} - \mathbf{I})\boldsymbol{\Omega}'.$$

Its latent roots are $\theta_1, \ldots, \theta_k$ together with $(p-k)$ unities. Hence

$$|\boldsymbol{\Sigma}_0|/|\boldsymbol{\Psi}| = |\boldsymbol{\Psi}^{-1/2}\boldsymbol{\Sigma}_0\boldsymbol{\Psi}^{-1/2}| = \theta_1\theta_2\ldots\theta_k$$

and

$$|\boldsymbol{\Sigma}_0| = (\theta_1\theta_2\ldots\theta_k)|\boldsymbol{\Psi}|.$$

We also have

$$|\mathbf{S}| = |\boldsymbol{\Psi}^{1/2}\mathbf{S}^*\boldsymbol{\Psi}^{1/2}| = |\mathbf{S}^*|\,|\boldsymbol{\Psi}| = (\theta_1\theta_2\ldots\theta_p)|\boldsymbol{\Psi}|.$$

Equation (4.5), with $\boldsymbol{\Lambda}_0$ and $\boldsymbol{\Sigma}_0$ in place of $\boldsymbol{\Lambda}$ and $\boldsymbol{\Sigma}$, may be put in the form

$$\mathbf{S}\boldsymbol{\Sigma}_0^{-1}\boldsymbol{\Lambda}_0 = \boldsymbol{\Lambda}_0.$$

Using this result we find that

$$\begin{aligned}\mathbf{S}\boldsymbol{\Sigma}_0^{-1} &= \mathbf{S}\boldsymbol{\Sigma}_0^{-1}(\boldsymbol{\Sigma}_0 - \boldsymbol{\Lambda}_0\boldsymbol{\Lambda}_0')\boldsymbol{\Psi}^{-1} \\ &= \mathbf{S}\boldsymbol{\Psi}^{-1} - (\mathbf{S}\boldsymbol{\Sigma}_0^{-1}\boldsymbol{\Lambda}_0)\boldsymbol{\Lambda}_0'\boldsymbol{\Psi}^{-1} \\ &= \mathbf{S}\boldsymbol{\Psi}^{-1} - \boldsymbol{\Lambda}_0\boldsymbol{\Lambda}_0'\boldsymbol{\Psi}^{-1}.\end{aligned}$$

Hence

$$\begin{aligned}\operatorname{tr}(\mathbf{S}\boldsymbol{\Sigma}_0^{-1}) &= \operatorname{tr}(\mathbf{S}\boldsymbol{\Psi}^{-1}) - \operatorname{tr}(\boldsymbol{\Lambda}_0\boldsymbol{\Lambda}_0'\boldsymbol{\Psi}^{-1}) \\ &= \operatorname{tr}(\boldsymbol{\Psi}^{-1/2}\mathbf{S}\boldsymbol{\Psi}^{-1/2}) - \operatorname{tr}(\boldsymbol{\Lambda}_0'\boldsymbol{\Psi}^{-1}\boldsymbol{\Lambda}_0) \\ &= \operatorname{tr}(\mathbf{S}^*) - \operatorname{tr}(\boldsymbol{\Theta} - \mathbf{I}) \\ &= (\theta_1 + \ldots + \theta_p) - [(\theta_1 - 1) + \ldots + (\theta_k - 1)] \\ &= (\theta_{k+1} + \ldots + \theta_p) + k. \end{aligned} \tag{4.15}$$

Substitution of the above expressions for $|\boldsymbol{\Sigma}_0|$, $|\mathbf{S}|$ and $\operatorname{tr}(\mathbf{S}\boldsymbol{\Sigma}_0^{-1})$ in the right-hand side of (4.14) finally gives

$$f_k(\boldsymbol{\Psi}) = \sum_{m=k+1}^{p} (\theta_m - \log_e \theta_m) - (p-k). \tag{4.16}$$

This expression for f_k shows that we were correct in making the elements of $\boldsymbol{\Theta}$ correspond to the k largest latent roots of $\mathbf{S}^*$. For suppose that in the right-hand side of (4.16) we exchange one of $\theta_{k+1}, \ldots, \theta_p$ for one of $\theta_1, \ldots, \theta_k$. The replaced root must exceed unity, as otherwise the corresponding column of $\mathbf{\Lambda}$ would acquire imaginary elements. But any such exchange would increase the value of f_k, since $x - \log_e x$ is an increasing function of x for $x > 1$.

An alternative expression for f_k in terms of the k largest latent roots of $\mathbf{S}^*$ and the elements of $\mathbf{\Psi}$ is

$$f_k(\mathbf{\Psi}) = \log(\theta_1\theta_2\ldots\theta_k) - (\theta_1 + \ldots + \theta_k) - \log|\mathbf{S}| + \log(\psi_1\psi_2\ldots\psi_p) + \operatorname{tr}(\mathbf{S}^*) - (p-k). \qquad (4.17)$$

In this formula note that

$$\operatorname{tr}(\mathbf{S}^*) = \sum_i (s_{ii}/\psi_i).$$

The second stage in the minimisation procedure consists in finding the $\mathbf{\Psi}$ that minimises $f(\mathbf{\Psi})$. For this we require the diagonal matrix $\partial f/\partial\mathbf{\Psi}$, whose diagonal elements are the derivatives $\partial f/\partial\psi_i$ $(i = 1, \ldots, p)$. This matrix is simply $\partial F/\partial\mathbf{\Psi}$ when evaluated with $\mathbf{\Lambda} = \mathbf{\Lambda}_0$. From (4.4) we have

$$\partial f/\partial\mathbf{\Psi} = \operatorname{diag}[\mathbf{\Sigma}_0^{-1}(\mathbf{\Sigma}_0 - \mathbf{S})\mathbf{\Sigma}_0^{-1}].$$

By use of (4.6) and (4.8) we find that

$$\mathbf{\Sigma}_0^{-1}(\mathbf{\Sigma}_0 - \mathbf{S})\mathbf{\Sigma}_0^{-1} = \mathbf{\Psi}^{-1}(\mathbf{\Sigma}_0 - \mathbf{S})\mathbf{\Psi}^{-1}.$$

Hence

$$\partial f/\partial\mathbf{\Psi} = \operatorname{diag}[\mathbf{\Psi}^{-1}(\mathbf{\Lambda}_0\mathbf{\Lambda}_0' + \mathbf{\Psi} - \mathbf{S})\mathbf{\Psi}^{-1}]. \qquad (4.18)$$

For the particular $\mathbf{\Psi}$ that minimises $f(\mathbf{\Psi})$ we have $\partial f/\partial\mathbf{\Psi} = \mathbf{0}$, and hence

$$\operatorname{diag}(\mathbf{\Lambda}_0\mathbf{\Lambda}_0' + \mathbf{\Psi} - \mathbf{S}) = \mathbf{0},$$

which is equivalent to

$$\mathbf{\Psi} = \operatorname{diag}(\mathbf{S} - \mathbf{\Lambda}_0\mathbf{\Lambda}_0'). \qquad (4.19)$$

This equation looks temptingly simple and earlier maximum likelihood procedures fell into the trap of attempting a direct solution of it. Unfortunately there is no guarantee that such a procedure will converge. It must be kept in mind that (4.19) is satisfied only at the minimum of f and that the elements of $\mathbf{\Lambda}_0$ are functions of the elements of $\mathbf{\Psi}$ as well as those of $\mathbf{S}$. The method that we describe in the next section makes no direct use of this equation. Instead it minimises f by evaluating at each step the function and its derivatives.

4.3 THE ITERATIVE PROCEDURE

The function f_k is minimised by the use of a method of Fletcher and Powell (1963), with a few modifications to make it suitable for our special purpose. The method is described in more detail in section A2.5 of Appendix II. The essence of the method is that in each iteration a second-degree approximation to the function f is used to estimate the minimum point. This results in a sequence of matrices $\mathbf{\Psi}^{(1)}, \mathbf{\Psi}^{(2)}, \ldots$, such that

$$f[\mathbf{\Psi}^{(s+1)}] < f[\mathbf{\Psi}^{(s)}].$$

The sequence converges rapidly to a final matrix of estimates $\hat{\mathbf{\Psi}}$. For each new $\mathbf{\Psi}^{(s)}$ we compute the corresponding $\mathbf{\Lambda}_0^{(s)}$ by the method of the previous section. Thus we have also a sequence of $\mathbf{\Lambda}$ matrices converging to a final matrix of estimates $\hat{\mathbf{\Lambda}}$.

In each iteration it is necessary to compute the value of f, as given by (4.17). We need also the values of the derivatives $\partial f/\partial \psi_i$. Equation (4.18) may be written, more conveniently for our present purpose, in the form

$$\partial f/\partial \mathbf{\Psi} = \text{diag}\,[\mathbf{\Psi}^{-1/2}\mathbf{\Omega}(\mathbf{\Theta}-\mathbf{I})\mathbf{\Omega}'\mathbf{\Psi}^{-1/2}+\mathbf{\Psi}^{-1}-\mathbf{\Psi}^{-1}\mathbf{S}\mathbf{\Psi}^{-1}],$$

from which we have

$$\partial f/\partial \psi_i = (1/\psi_i)\left[\sum_{r=1}^{k}(\theta_r-1)\omega_{ir}^2+1-s_{ii}/\psi_i\right], \qquad (4.20)$$

where ω_{ir} is the ith element of the vector $\boldsymbol{\omega}_r$. The computation of the derivatives thus requires the numerical evaluation of the first k latent roots and vectors of the symmetric matrix $\mathbf{S}^*$.

Since a second-degree approximation to the function f is required in the Fletcher and Powell method, it is necessary to compute in each iteration a symmetric and positive definite matrix $\mathbf{E}$, of order p. This results in a sequence of matrices $\mathbf{E}^{(1)}, \mathbf{E}^{(2)}, \ldots$, that converges to the inverse of the matrix of second-order partial derivatives $\partial^2 f/\partial\psi_i\partial\psi_j$ evaluated at the minimum. As we shall show in section 4.5, it is possible to find approximations to these second derivatives, and this enables us to compute a very good initial $\mathbf{E}$ matrix.

To start the iterative procedure we require an initial $\mathbf{\Psi}$. We could take $\mathbf{\Psi}^{(1)} = \text{diag}\,(\mathbf{S})$ or, in cases where the variates are standardised $\mathbf{\Psi}^{(1)} = \mathbf{I}$. A better choice for $\mathbf{\Psi}^{(1)}$, requiring fewer iterations, is however given by

$$\psi_i^{(1)} = (1-\tfrac{1}{2}k/p)(1/s^{ii}) \qquad (i = 1, \ldots, p), \qquad (4.21)$$

where s^{ij} denotes the element in the ith row and jth column of $\mathbf{S}^{-1}$. This choice has been justified by Jöreskog (1963) and appears to work reasonably well in practice.

A difficulty that has beset all methods of finding maximum likelihood estimates is that the likelihood function may not have any true maximum, in the sense that all first derivatives vanish, within the region for which all the ψ_i are positive. If this happens, it may be due to the fact that the model is inappropriate. But this is not necessarily the case, especially when the sample size is not very large. As Browne (1965) discovered when analysing the results of artificial sampling experiments, the trouble may arise simply from sampling errors. In such cases it is found that one or more elements of $\mathbf{\Psi}$ tend to zero in the course of iteration and would become negative if allowed to do so.

To overcome this difficulty we find the minimum of f within the region R_ε for which $\psi_i \geqslant \varepsilon$ for all values of i, where ε is an arbitrarily small positive value. In practice, for standardised variates, ε has usually been taken as 0·005. If the minimum of f is found in the interior of R_ε, the set of maximum likelihood estimates is termed a *proper solution*. If, on the other hand, the minimum occurs on the boundary of R_ε, the set is termed an *improper solution*. Suppose that in the course of iteration a particular ψ_i attains the value ε and that the derivative $\partial f/\partial\psi_i$ is positive. Then any increase in ψ_i would increase the value of f, while any decrease would mean going outside R_ε. So in subsequent iterations ψ_i is fixed at the value ε. It is possible that $\partial f/\partial\psi_i$ might finally become negative, in which case ψ_i would be free to increase; but this does not normally happen.

In the literature of factor analysis improper solutions have usually been known as Heywood cases. What is surprising is the frequency with which they appear to occur. In a study conducted by Mattsson, Olsson and Rosén (1966), some of the results of which have been summarised by Jöreskog (1967), eleven sets of data were analysed by the method of this chapter. In all but two sets improper solutions were obtained for some values of k. In some cases these improper solutions had not been suspected when the data were originally analysed, due to the fact that the iterative procedure then used was stopped too soon.

Suppose that an improper solution is obtained in which $m(\leq k)$ of the elements of $\mathbf{\Psi}$, say $\psi_1, \ldots, \psi_m$, are estimated as ε. The hypothesis is then reframed in terms of the model of section 2.4 of Chapter 2. We make the assumption that the true values of $\psi_1, \ldots, \psi_m$ are all zero. This is open to objection, since we are not strictly justified in altering the model and adopting a different hypothesis in the course of analysing one set of data. But to some extent the method that we have described may be regarded as exploratory in character, as we shall subsequently make clear.

Assuming that $\psi_1, \ldots, \psi_m$ are all zero, we return to the model and notation of section 2.4. Let $\mathbf{S}$ be partitioned in the same way as $\mathbf{\Sigma}$. Then we have

$$\mathbf{S} = \begin{bmatrix} \mathbf{S}_{11} & \mathbf{S}_{12} \\ \mathbf{S}_{21} & \mathbf{S}_{22} \end{bmatrix}.$$

When the first m variates $\mathbf{x}_1$ have been eliminated the sample partial covariance matrix $\mathbf{S}_{22\cdot 1}$ for the remaining variates $\mathbf{x}_2$ may be defined by

$$\mathbf{S}_{22\cdot 1} = \mathbf{S}_{22} - \mathbf{S}_{21}\mathbf{S}_{11}^{-1}\mathbf{S}_{12}.$$

This matrix is of order $p-m$ and contains all the available information about $\mathbf{\Sigma}_{22\cdot 1}$. The elements of $n\mathbf{S}_{22\cdot 1}$ follow a Wishart distribution with $n-m$ degrees of freedom. In the same notation as before the distribution is $W(\mathbf{\Sigma}_{22\cdot 1}, n-m)$. As in section 2.4, the first m factors $\mathbf{f}_1$ are chosen as a linear transformation of $\mathbf{x}_1$. When $m < k$, the remaining factors are assumed to be responsible for all correlation in $\mathbf{S}_{22\cdot 1}$. If $m = k$, there should be no significant correlation in $S_{22\cdot 1}$.

If $m = 1$, the maximum likelihood estimate $\hat{\mathbf{\Lambda}}_{11}$ of $\mathbf{\Lambda}_{11}$ is merely the scalar $\sqrt{s_{11}}$. Otherwise it is obtained by performing a principal component analysis on $\mathbf{S}_{11}$ and it satisfies

$$\hat{\mathbf{\Lambda}}_{11}\hat{\mathbf{\Lambda}}_{11}' = \mathbf{S}_{11}. \tag{4.22}$$

The matrix $\mathbf{\Sigma}_{21}\mathbf{\Sigma}_{11}^{-1}$ has as its elements the linear regression coefficients of $\mathbf{x}_2$ on $\mathbf{x}_1$. By standard regression theory its maximum likelihood estimate is $\mathbf{S}_{21}\mathbf{S}_{11}^{-1}$. Hence the maximum likelihood estimate of $\mathbf{\Lambda}_{21} = \mathbf{\Sigma}_{21}\mathbf{\Sigma}_{11}^{-1}\mathbf{\Lambda}_{11}$ is given by

$$\hat{\mathbf{\Lambda}}_{21} = \mathbf{S}_{21}\mathbf{S}_{11}^{-1}\hat{\mathbf{\Lambda}}_{11}. \tag{4.23}$$

Finally we estimate $\mathbf{\Psi}_2$ and $\mathbf{\Lambda}_{22}$, when $m < k$, by applying the maximum likelihood method to the matrix $\mathbf{S}_{22\cdot 1}$. The procedure is exactly as before except that n, p and k are replaced respectively by $n-m$, $p-m$ and $k-m$.

A satisfactory feature of the maximum likelihood method is that it is independent of the scales of measurement of the variates. This is clear from an examination of the various equations of estimation. The iterative procedure depends only on finding the first k latent roots and vectors of the matrix $\mathbf{S}^* = \mathbf{\Psi}^{-1/2}\mathbf{S}\mathbf{\Psi}^{-1/2}$, which is independent of the metric of $\mathbf{x}$. This fact is of great computational convenience, since it means that in performing numerical calculations we can standardise all variates and substitute for $\mathbf{S}$ the sample correlation matrix of $\mathbf{x}$, having unities as its diagonal elements.

A simple interpretation can be given of the elements of the diagonal matrix $\mathbf{\Lambda}'\mathbf{\Psi}^{-1}\mathbf{\Lambda}$ that has frequently occurred in the course of the algebra. If the variates are rescaled so that the residual variance of each is unity, then λ_{ir}^2/ψ_i represents the part of the variance of x_i that is due to the rth factor. If we consider all p variates, the total variance in $\mathbf{x}$ due to the rth factor is $\sum_i (\lambda_{ir}^2/\psi_i)$. This is the rth diagonal element of $\mathbf{\Lambda}'\mathbf{\Psi}^{-1}\mathbf{\Lambda}$. Our choice of factors is such that the first one makes a maximum contribution to the variance in $\mathbf{x}$, the second makes a maximum contribution subject to being uncorrelated with the first, and so on.

A difficulty that occasionally arises with any application of the maximum likelihood method is that the likelihood function may have more than one maximum within the parameter space. In most problems this happens mainly with small samples; but all we can say with assurance is that if the model chosen is the appropriate one, the probability of the equations of estimation having a unique solution tends to unity as the sample size $N \to \infty$. In our problem it would be possible to detect multiple solutions only by performing a number of different minimisations of $f_k(\mathbf{\Psi})$, starting each time with a different initial $\mathbf{\Psi}$. When multiple solutions occur the preferred one is that corresponding to the absolute minimum of f_k. An alternative minimisation of f_k was discovered by Jöreskog (1967) in one of the 11 sets of data already mentioned and in others by Tumura, Fukutomi and Assoo (1968). In each case the two alternative minimisations corresponded to improper solutions having different elements of $\mathbf{\Psi}$ on the boundary of R_ε. Evidently extreme caution should be used in attempting to interpret data from which an improper solution has been obtained, and the existence of alternative minimisations should be suspected.

4.4 TESTS OF HYPOTHESES

One of the main advantages of using the maximum likelihood method of estimation is that it enables us to test the hypothesis H_k that, for specified k, there are k common factors, under the assumptions made at the beginning of this chapter. For this purpose we make use of the appropriate likelihood ratio criterion. We assume that a proper solution has been obtained.

Let Ω denote the set of all matrices $\mathbf{\Sigma}$ that are of order p, symmetric and positive definite. Let ω be the subset for which equation (2.2) holds, in accordance with H_k. Let L_Ω and L_ω denote the maxima of L in Ω and ω respectively. It is well known in multivariate analysis that L attains its maximum in Ω when $\mathbf{\Sigma} = \mathbf{S}$. Hence from the expression (4.1) for $\log L$ we have

$$\log L_\Omega = -\tfrac{1}{2}n\,(\log|\mathbf{S}|+p). \tag{4.24}$$

To find L_ω we substitute $\hat{\mathbf{\Sigma}}$ for $\mathbf{\Sigma}$ in (4.1), where

$$\hat{\mathbf{\Sigma}} = \hat{\mathbf{\Lambda}}\hat{\mathbf{\Lambda}}' + \hat{\mathbf{\Psi}}$$

and where $\hat{\mathbf{\Lambda}}$ and $\hat{\mathbf{\Psi}}$ are the maximum likelihood estimates found previously. Hence

$$\log L_\omega = -\tfrac{1}{2}n\,[\log|\hat{\mathbf{\Sigma}}| + \operatorname{tr}(\mathbf{S}\hat{\mathbf{\Sigma}}^{-1})]. \tag{4.25}$$

The likelihood ratio is $\lambda = L_\omega/L_\Omega$. Though we cannot in this case find its exact distribution, it is known that for large samples $-2\log_e\lambda$ is distributed approximately as χ^2 if H_k is true. The number of degrees of freedom for χ^2 is the number of parameters in Ω, that is $\frac{1}{2}p(p+1)$, minus the number of free parameters in ω, i.e. under H_k. The number of degrees of freedom is thus the value of s calculated in section 2.3 of Chapter 2 and is $\frac{1}{2}[(p-k)^2-(p+k)]$. It is positive if H_k is non-trivial.

From (4.24) and (4.25) we have

$$\begin{aligned} -2\log\lambda &= -2\log L_\omega + 2\log L_\Omega \\ &= n\,[\log|\hat{\mathbf{\Sigma}}| + \operatorname{tr}(\mathbf{S}\hat{\mathbf{\Sigma}}^{-1}) - \log|\mathbf{S}| - p]. \end{aligned} \tag{4.26}$$

On comparing this with equation (4.14) it is clear that $-2\log\lambda$ is simply n times $f_k(\mathbf{\Psi})$ evaluated at the minimum. Thus

$$-2\log\lambda = nf_k(\hat{\mathbf{\Psi}}) = nF_k(\hat{\mathbf{\Lambda}}, \hat{\mathbf{\Psi}}).$$

This explains our choice of F_k.

The numerical value of $f_k(\hat{\mathbf{\Psi}})$ is determined by the method of the previous section. As we shall show, there are various algebraic expressions for it that are of interest. By use of (4.6) and (4.8) with the appropriate circumflex accents inserted, we find that

$$\mathbf{S}\hat{\mathbf{\Sigma}}^{-1} = \mathbf{I} - (\hat{\mathbf{\Sigma}} - \mathbf{S})\hat{\mathbf{\Sigma}}^{-1} = \mathbf{I} - (\hat{\mathbf{\Sigma}} - \mathbf{S})\hat{\mathbf{\Psi}}^{-1}. \tag{4.27}$$

Since, by (4.19), $\hat{\mathbf{\Sigma}} - \mathbf{S}$ has zero diagonal elements, it follows that

$$\operatorname{tr}(\mathbf{S}\hat{\mathbf{\Sigma}}^{-1}) = \operatorname{tr}(\mathbf{I}_p) = p. \tag{4.28}$$

Use of this result in (4.14), with $\mathbf{\Psi}$ and $\mathbf{\Lambda}_0$ replaced by $\hat{\mathbf{\Psi}}$ and $\hat{\mathbf{\Lambda}}$, gives

$$f_k(\hat{\mathbf{\Psi}}) = \log|\hat{\mathbf{\Sigma}}| - \log|\mathbf{S}|. \tag{4.29}$$

In the trivial case where $k = 0$ we have $\hat{\mathbf{\Sigma}} = \operatorname{diag}(\mathbf{S})$. Hence the likelihood ratio criterion for testing independence of the variates is then

$$-n\log(|\mathbf{S}|/|\hat{\mathbf{\Sigma}}|) = -n\log|\mathbf{R}|,$$

where $\mathbf{R}$ is the sample correlation matrix. It was shown by Box (1949) that the χ^2 approximation is improved if in this criterion

n is replaced by $n-(2p+5)/6$. In view of this Bartlett (1951) suggested that for $k > 0$ an improvement would be obtained by the use of Box's multiplying factor with n and p replaced respectively by $n-k$ and $p-k$. Though this is to some extent conjectural, we shall accordingly test H_k by means of the criterion

$$U_k = [n-(2p+5)/6-2k/3]f_k(\hat{\mathbf{\Psi}}). \tag{4.30}$$

Under H_k this is distributed, for moderately large n, very nearly as χ^2 with its number of degrees of freedom given by

$$d_k = \tfrac{1}{2}[(p-k)^2-(p+k)]. \tag{4.31}$$

The χ^2 approximation can probably be trusted if $n-p \geqslant 50$.

Let $\hat{\theta}_m$ denote the mth latent root of $\hat{\mathbf{\Psi}}^{-1/2}\mathbf{S}\hat{\mathbf{\Psi}}^{-1/2}$. Then from equations (4.28) and (4.15), with $\mathbf{\Sigma}_0$ and θ_m replaced respectively by $\hat{\mathbf{\Sigma}}$ and $\hat{\theta}_m$, we have

$$\hat{\theta}_{k+1}+\ldots+\hat{\theta}_p = \text{tr}\,(\mathbf{S}\hat{\mathbf{\Sigma}}^{-1})-k = p-k. \tag{4.32}$$

Use of this result in (4.16), with insertion of the appropriate circumflex accents, gives

$$f_k(\hat{\mathbf{\Psi}}) = -\sum_{m=k+1}^{p} (\log_e \hat{\theta}_m). \tag{4.33}$$

When n is large the values of $(\hat{\theta}_m-1)$ for $m = k+1, \ldots, p$ are small in probability. Hence, by expanding $\log_e \hat{\theta}_m$ in powers of $(\hat{\theta}_m-1)$ and using (4.32), we have

$$f_k(\hat{\mathbf{\Psi}}) \approx \tfrac{1}{2}[(\hat{\theta}_{k+1}-1)^2+\ldots+(\hat{\theta}_p-1)^2]. \tag{4.34}$$

Thus the criterion U_k is a measure of how much the roots $\hat{\theta}_{k+1}, \ldots, \hat{\theta}_p$ differ from unity.

We derive one other approximation to $f_k(\hat{\mathbf{\Psi}})$. From (4.29) and (4.27) we have

$$f_k(\hat{\mathbf{\Psi}}) = -\log|\mathbf{S}\hat{\mathbf{\Sigma}}^{-1}| = -\log|\mathbf{I}+\mathbf{X}|,$$

where

$$\mathbf{X} = (\mathbf{S}-\hat{\mathbf{\Sigma}})\hat{\mathbf{\Psi}}^{-1}.$$

The diagonal elements of $\mathbf{X}$ are zero and the other elements are small in probability. Hence, denoting these elements by x_{ij}, we have

$$|\mathbf{I}+\mathbf{X}| \approx 1-\sum_{i<j}(x_{ij}x_{ji})$$

and

$$f_k(\hat{\mathbf{\Psi}}) \approx \sum_{i<j}(x_{ij}x_{ji}) = \sum_{i<j}[(s_{ij}-\hat{\sigma}_{ij})^2/(\hat{\psi}_i\hat{\psi}_j)]. \tag{4.35}$$

In this expression $s_{ij}-\hat{\sigma}_{ij}$ represents the residual covariance of x_i and x_j after eliminating k common factors. The criterion U_k is therefore a measure of how much the residual covariances differ from zero. Before the advent of computers the above expression was often used as a method of approximating U_k.

In the above derivation of the test criterion for H_k we assumed that a proper solution had been obtained. Now suppose that an improper solution is found, having m elements of $\mathbf{\Psi}$ estimated as ε. Instead of calculating U_k as above, we first eliminate the corresponding m variates $\mathbf{x}_1$ and obtain the proper solution based on $\mathbf{S}_{22\cdot1}$. The test criterion is then found as before with $n-m$, $p-m$ and $k-m$ in place of n, p and k respectively. Note that the number of degrees of freedom for χ^2 is increased by m. This is because there are m fewer parameters in ω than before. The multiplying factor in (4.30) is, however, unchanged. The criterion is now used to test the hypothesis that the $p-m$ remaining variates $\mathbf{x}_2$ depend upon $k-m$ common factors. If $m = k$, then $\mathbf{S}_{22\cdot1}$ is tested for independence.

Throughout the foregoing discussions we have supposed that k can be specified in advance, though in most exploratory factor studies this is not in fact possible. We have also expressed the problem in terms of the classical theory of hypothesis testing. In this theory if U_k exceeds the significance level for χ^2 with d_k degrees of freedom, corresponding to an arbitrary significance probability P, we reject H_k; otherwise H_k is accepted. In practice the problem is usually not that of testing a particular hypothesis but rather of deciding how many common factors it is worth fitting to the data.

To cope with this problem a sequential procedure for determining k is used. Starting with some small value k_1, which can be taken as 1, or even 0, we estimate the parameters with $k = k_1$. Having reached a proper solution, possibly with the elimination of some variates, we calculate the test criterion. If this is not significant at the chosen level, we accept H_k with $k = k_1$. But if the test criterion is significant, we next try $k = k_1+1$. The procedure continues, with k increasing by 1 at each step, until either H_k is accepted for some k or H_k is rejected with $k = k_2$, where k_2 is the largest value of k for which $d_k > 0$. In the latter case no non-trivial H_k is acceptable.

The above procedure is open to objection on the grounds that the significance level for the test criterion has not been adjusted to take into account the fact that a sequence of hypotheses is being tested, with each one dependent on the rejection of all predecessors. The mathematical complications involved make such an adjustment almost impossible to carry out. The use of the significance level of χ^2 at each step seems unlikely to cause serious error in practice. If the model with k^* as the true number of factors is correct, there is a small probability that the value of k determined by the sequential process will be less than k^*. The probability of k being greater than k^* cannot exceed the nominal significance probability P and will usually be only slightly less than P.

If the hypothesis H_k is accepted, this does not prove that k is the true number of common factors present. All it means is that there would be no point in fitting further factors to the data, since these would be indistinguishable from experimental error. On the other hand, if the final value of k is at all large, it may be difficult to interpret all the factors, even after rotation or transformation. Particularly with large samples, the last few factors, though significant in the statistical sense, are often unimportant in that they contribute little either to the correlation or to the variance in $\mathbf{x}$. It should always be kept firmly in mind that, except in artificial sampling experiments, the basic factor model is, like other models, useful only as an approximation to reality, and it should not be taken too seriously.

4.5 APPROXIMATIONS TO SECOND ORDER DERIVATIVES

When discussing the application of Fletcher and Powell's method of function minimisation in section 4.3, we mentioned a sequence of matrices $\mathbf{E}^{(1)}, \mathbf{E}^{(2)}, \ldots$, that converged to the inverse of the matrix of second-order derivatives $\partial^2 f/\partial\psi_i\partial\psi_j$, evaluated at the minimum of f. We now show how these derivatives may be approximated, thereby providing a good initial matrix $\mathbf{E}^{(1)}$.

For convenience we modify the notation of previous sections. In future $\mathbf{\Lambda}$, $\mathbf{\Psi}$, and $\mathbf{\Sigma}$ will refer to true values of the parameters. The diagonal matrix $\mathbf{\Theta}$ has as its elements the first k latent roots of $\mathbf{\Psi}^{-1/2}\mathbf{\Sigma}\mathbf{\Psi}^{-1/2}$ and not, as formerly, those of $\mathbf{\Psi}^{-1/2}\mathbf{S}\mathbf{\Psi}^{-1/2}$. The columns of $\mathbf{\Omega}$ are the corresponding standardised latent vectors. Thus the elements of $\mathbf{\Theta}$ and $\mathbf{\Omega}$ are now functions of the true values of the parameters only. With this notation we have

$$\mathbf{\Lambda} = \mathbf{\Psi}^{1/2}\mathbf{\Omega}(\mathbf{\Theta}-\mathbf{I})^{1/2} \tag{4.36}$$

and

$$\mathbf{\Lambda}'\mathbf{\Psi}^{-1}\mathbf{\Lambda} = \mathbf{\Theta}-\mathbf{I}. \tag{4.37}$$

We shall correspondingly use $\hat{\mathbf{\Theta}}_0$ and $\hat{\mathbf{\Omega}}_0$ to refer to the first k latent roots and vectors of $\mathbf{\Psi}^{-1/2}\mathbf{S}\mathbf{\Psi}^{-1/2}$. The maximum likelihood estimate of $\mathbf{\Lambda}$ when $\mathbf{\Psi}$ is known is now denoted by $\hat{\mathbf{\Lambda}}_0$. Hence equations (4.36) and (4.37) are satisfied with $\hat{\mathbf{\Lambda}}_0$, $\hat{\mathbf{\Omega}}_0$ and $\hat{\mathbf{\Theta}}_0$ in the place of $\mathbf{\Lambda}, \mathbf{\Omega}$ and $\mathbf{\Theta}$ respectively. The symbols $\hat{\mathbf{\Theta}}$ and $\hat{\mathbf{\Omega}}$, with circumflex accents but without zero suffices, refer to the first k latent roots and vectors of $\hat{\mathbf{\Psi}}^{-1/2}\mathbf{S}\hat{\mathbf{\Psi}}^{-1/2}$, where $\hat{\mathbf{\Psi}}$ is the maximum likelihood estimate of $\mathbf{\Psi}$. The corresponding maximum likelihood estimate of $\mathbf{\Lambda}$ when $\mathbf{\Psi}$ is unknown and estimated by $\hat{\mathbf{\Psi}}$ is, as before, denoted by $\hat{\mathbf{\Lambda}}$. So equations (4.36) and (4.37) are satisfied when $\mathbf{\Lambda}$, $\mathbf{\Psi}$, $\mathbf{\Theta}$ and $\mathbf{\Omega}$ all receive circumflex accents.

At this stage some care is needed to establish a correspondence between the elements of $\mathbf{\Lambda}$, $\hat{\mathbf{\Lambda}}$ and $\hat{\mathbf{\Lambda}}_0$. In each of the matrices $\mathbf{\Theta}$, $\hat{\mathbf{\Theta}}$ and $\hat{\mathbf{\Theta}}_0$ the diagonal elements are assumed to be arranged in decreasing order of magnitude. With small samples this does not ensure complete correspondence between the columns of the various $\mathbf{\Omega}$ matrices. A column of $\hat{\mathbf{\Omega}}$ or of $\hat{\mathbf{\Omega}}_0$ is sometimes found to resemble not the corresponding column of $\mathbf{\Omega}$ but an adjacent one. This difficulty is most likely to arise if two elements of $\mathbf{\Theta}$ are close to each other. However, when $n \to \infty$ each element of $\hat{\mathbf{\Theta}}$ or of $\hat{\mathbf{\Theta}}_0$ tends in probability to the corresponding element of $\mathbf{\Theta}$. We therefore assume that the value of n is sufficiently large for sampling errors in the elements of $\hat{\mathbf{\Theta}}$ and $\hat{\mathbf{\Theta}}_0$ to be small in probability compared with the differences between successive elements of $\mathbf{\Theta}$. We can then set up an almost certain correspondence between the columns of $\hat{\mathbf{\Omega}}, \hat{\mathbf{\Omega}}_0$ and $\mathbf{\Omega}$.

A consistent choice of sign must be made for the columns of the various $\mathbf{\Omega}$ matrices. For each column we may, for example, decide that a specified element in it is to be made positive; but whatever rule is adopted, it must be the same for $\hat{\mathbf{\Omega}}$, $\hat{\mathbf{\Omega}}_0$ and $\mathbf{\Omega}$. Then each element of $\hat{\mathbf{\Lambda}}$ or of $\hat{\mathbf{\Lambda}}_0$ is a consistent estimate of the corresponding element of $\mathbf{\Lambda}$.

The first-order derivatives $\partial f/\partial\psi_i$ are the diagonal elements of the matrix $\partial f/\partial\mathbf{\Psi}$, as given by equation (4.18). In our present notation this equation is

$$\partial f/\partial\mathbf{\Psi} = \text{diag}\,[\mathbf{\Psi}^{-1}(\hat{\mathbf{\Lambda}}_0\hat{\mathbf{\Lambda}}_0'+\mathbf{\Psi}-\mathbf{S})\mathbf{\Psi}^{-1}]. \tag{4.38}$$

In the right-hand side of this the elements of $\hat{\mathbf{\Lambda}}_0$ are complicated functions of the ψ_j. We therefore express it approximately in terms of the matrix $\mathbf{\Lambda}\mathbf{\Lambda}'+\mathbf{\Psi}-\mathbf{S}$.

Let us define the matrix $\mathbf{Z}$ by

$$\mathbf{Z} = \hat{\mathbf{\Lambda}}_0-\mathbf{\Lambda}.$$

As it is assumed throughout that n is reasonably large, the elements of $\mathbf{Z}$, which represent sampling errors, are small in probability. We therefore neglect terms of the second degree in the elements of $\mathbf{Z}$, and in the ensuing algebra all equations are correct to this order of approximation. Thus we have

$$\hat{\mathbf{\Lambda}}_0\hat{\mathbf{\Lambda}}_0' = (\mathbf{\Lambda}+\mathbf{Z})(\mathbf{\Lambda}+\mathbf{Z})' = \mathbf{\Lambda}\mathbf{\Lambda}'+\mathbf{\Lambda}\mathbf{Z}'+\mathbf{Z}\mathbf{\Lambda}'. \tag{4.39}$$

In our present notation equation (4.8) is equivalent to

$$(\hat{\mathbf{\Sigma}}_0-\mathbf{S})\mathbf{\Psi}^{-1}\hat{\mathbf{\Lambda}}_0 = \mathbf{0},$$

where

$$\hat{\mathbf{\Sigma}}_0 = \hat{\mathbf{\Lambda}}_0\hat{\mathbf{\Lambda}}_0'+\mathbf{\Psi}.$$

Since the elements of $(\hat{\mathbf{\Sigma}}_0-\mathbf{S})$ are small in probability, this is asymptotically equivalent to

$$(\hat{\mathbf{\Sigma}}_0-\mathbf{S})\mathbf{\Psi}^{-1}\mathbf{\Lambda} = \mathbf{0}. \tag{4.40}$$

From (4.39) we have

$$\hat{\mathbf{\Sigma}}_0 = \mathbf{\Sigma}+\mathbf{\Lambda}\mathbf{Z}'+\mathbf{Z}\mathbf{\Lambda}'. \tag{4.41}$$

We now define the symmetric matrix $\mathbf{\Phi} = [\phi_{ij}]$, of order p, by

$$\mathbf{\Phi} = \mathbf{\Psi}^{-1/2}(\mathbf{I}-\mathbf{\Omega}\mathbf{\Omega}')\mathbf{\Psi}^{-1/2}, \tag{4.42}$$

which is the same as

$$\mathbf{\Phi} = \mathbf{\Psi}^{-1}-\mathbf{\Psi}^{-1}\mathbf{\Lambda}(\mathbf{\Theta}-\mathbf{I})^{-1}\mathbf{\Lambda}'\mathbf{\Psi}^{-1}. \tag{4.43}$$

Using (4.37) it is easy to verify that

$$\mathbf{\Phi}\mathbf{\Lambda} = \mathbf{0}, \tag{4.44}$$

and from (4.40) and (4.43) it follows that

$$(\hat{\mathbf{\Sigma}}_0-\mathbf{S})\mathbf{\Phi} = (\hat{\mathbf{\Sigma}}_0-\mathbf{S})\mathbf{\Psi}^{-1}. \tag{4.45}$$

Hence, by use of (4.45), (4.41) and (4.44), we have

$$\begin{aligned}\mathbf{\Psi}^{-1}(\hat{\mathbf{\Sigma}}_0-\mathbf{S})\mathbf{\Psi}^{-1} &= \mathbf{\Phi}(\hat{\mathbf{\Sigma}}_0-\mathbf{S})\mathbf{\Phi}\\ &= \mathbf{\Phi}(\mathbf{\Sigma}+\mathbf{\Lambda}\mathbf{Z}'+\mathbf{Z}\mathbf{\Lambda}'-\mathbf{S})\mathbf{\Phi}\\ &= \mathbf{\Phi}(\mathbf{\Sigma}-\mathbf{S})\mathbf{\Phi}.\end{aligned} \tag{4.46}$$

Thus equation (4.38) is asymptotically equivalent to

$$\partial f/\partial\mathbf{\Psi} = \operatorname{diag}\,[\mathbf{\Phi}(\mathbf{\Sigma}-\mathbf{S})\mathbf{\Phi}],$$

from which we have

$$\partial f/\partial\psi_i = [\mathbf{\Phi}(\mathbf{\Sigma}-\mathbf{S})\mathbf{\Phi}]_{ii}. \tag{4.47}$$

Differentiating with respect to ψ_j, we now find that

$$\partial^2 f/\partial\psi_i\partial\psi_j = [\mathbf{\Phi}(\partial\mathbf{\Sigma}/\partial\psi_j)\mathbf{\Phi}]_{ii}.$$

The terms that would arise on the right-hand side of this equation from the differentiation of $\mathbf{\Phi}$ have been neglected, since the elements of $\mathbf{\Sigma}-\mathbf{S}$ are small in probability. Since $E(\mathbf{\Sigma}-\mathbf{S}) = \mathbf{0}$, we may alternatively use the above expression as an approximation to $E(\partial^2 f/\partial\psi_i\partial\psi_j)$. As all elements of $\partial\mathbf{\Sigma}/\partial\psi_j$ are zero except for the jth diagonal element, which is unity, we have, asymptotically,

$$\partial^2 f/\partial\psi_i\partial\psi_j = E(\partial^2 f/\partial\psi_i\partial\psi_j) = \phi_{ij}^2. \tag{4.48}$$

Let the symmetric matrix $\mathbf{G} = [g_{ij}]$ be defined by

$$g_{ij} = \phi_{ij}^2 \qquad (i, j = 1, \ldots, p). \tag{4.49}$$

We shall prove that, in general, $\mathbf{G}$ is positive definite. We do this by showing that the quadratic form

$$Q = \sum_i \sum_j g_{ij} v_i v_j$$

is positive for any set of values $v_1, \ldots, v_p$ that are not all zero.

Since $\mathbf{\Omega}'\mathbf{\Omega} = \mathbf{I}$, we can find an orthogonal matrix of order p whose first k columns form $\mathbf{\Omega}$. Denoting the part consisting of the remaining $(p-k)$ columns by $\mathbf{\Omega}_1$, we have

$$\mathbf{I} - \mathbf{\Omega}\mathbf{\Omega}' = \mathbf{\Omega}_1\mathbf{\Omega}_1'.$$

Hence from the definition (4.42) of $\mathbf{\Phi}$ we may write

$$\mathbf{\Phi} = \mathbf{\Psi}^{-1/2}\mathbf{\Omega}_1\mathbf{\Omega}_1'\mathbf{\Psi}^{-1/2} = \mathbf{\Gamma}\mathbf{\Gamma}',$$

where

$$\mathbf{\Gamma} = \mathbf{\Psi}^{-1/2}\mathbf{\Omega}_1.$$

Let $\mathbf{V}$ be the diagonal matrix whose diagonal elements are $v_1, \ldots, v_p$. Then for the quadratic form Q we have

$$\begin{aligned} Q &= \sum_i \sum_j [(\phi_{ij} v_j)(\phi_{ji} v_i)] \\ &= \operatorname{tr}[(\mathbf{\Phi}\mathbf{V})(\mathbf{\Phi}\mathbf{V})] \\ &= \operatorname{tr}[(\mathbf{\Gamma}\mathbf{\Gamma}'\mathbf{V})^2] \\ &= \operatorname{tr}[(\mathbf{\Gamma}'\mathbf{V}\mathbf{\Gamma})^2]. \end{aligned}$$

The last expression given above is the same as the sum of squares of elements of $\mathbf{\Gamma}'\mathbf{V}\mathbf{\Gamma}$. Hence Q cannot be negative. It is zero if and only if this matrix is null. Since $\mathbf{\Gamma}'\mathbf{V}\mathbf{\Gamma}$ is symmetric and of order $p-k$, this condition may be regarded as a set of $\frac{1}{2}(p-k)(p-k+1)$ homogeneous linear equations for the p unknowns $v_1, \ldots, v_p$. Assuming that these equations are linearly independent, they can be satisfied by values of the v_i that are not all zero only if

$$\tfrac{1}{2}(p-k)(p-k+1) < p,$$

that is, only if

$$d_k = \tfrac{1}{2}[(p-k)^2 - (p+k)] < 0.$$

But for a non-trivial hypothesis H_k we have $d_k > 0$, and so Q must be positive. Hence the matrix $\mathbf{G}$ is positive definite and has a positive definite inverse $\mathbf{G}^{-1}$.

In the foregoing argument $\mathbf{\Psi}$ and $\mathbf{\Lambda}$ denoted true values of the parameters. In the Fletcher and Powell method, however, we need to approximate the second-order derivatives $\partial^2 f/\partial\psi_i\partial\psi_j$ for some trial matrix $\mathbf{\Psi}$ and with $\mathbf{\Lambda}$ satisfying (4.9). We may still use equation (4.48) for this purpose provided that the matrix $(\mathbf{\Lambda\Lambda}'+\mathbf{\Psi})$ is reasonably close to $\mathbf{S}$, which is likely to be the case if H_k is true and if $\mathbf{\Psi}$ is not too far from the maximum likelihood estimate $\hat{\mathbf{\Psi}}$. For the approximations that were made in deriving (4.48) depend essentially upon the assumption that the elements of $\mathbf{\Sigma}-\mathbf{S}$ are small. In evaluating the matrix $\mathbf{\Phi}$ by means of (4.42) we use the latent vectors corresponding to the first k latent roots of $\mathbf{\Psi}^{-1/2}\mathbf{S}\mathbf{\Psi}^{-1/2}$, for the trial $\mathbf{\Psi}$, as estimates of the columns of $\mathbf{\Omega}$. The matrix $\mathbf{G}$, given by (4.49), is inverted to provide the initial $\mathbf{E}$ matrix for the Fletcher and Powell method.

The initial $\mathbf{\Psi}$ matrix given by (4.21) is as a rule not close enough to $\hat{\mathbf{\Psi}}$ for good approximations to $\mathbf{\Phi}$ and $\mathbf{G}$ to be obtained. Experience has shown that, with this $\mathbf{\Psi}^{(1)}$, it is best when minimising the function f to start with two steepest descent iterations. The method of steepest descent (see Appendix II), though ineffective in the neighbourhood of the minimum, works very well at the beginning of the minimisation. The $\mathbf{\Psi}$ that is obtained after two iterations is usually very much nearer than $\mathbf{\Psi}^{(1)}$ to $\hat{\mathbf{\Psi}}$. From it we find $\mathbf{\Phi}$, $\mathbf{G}$ and $\mathbf{G}^{-1}$, the last of which provides a good initial $\mathbf{E}$ matrix.

4.6 NUMERICAL EXAMPLES

Two numerical examples are used to illustrate the methods previously described. The analyses that we give were carried out by Jöreskog (1967). In the first example only proper solutions were obtained. In the second, improper solutions were found for some values of k.

The first set of data was originally analysed by Emmett (1949). It consists of a sample of 211 observations of 9 variates. Thus $n = 210$ and $p = 9$. The standardised covariance matrix is given in *Table 4.1*. As with other symmetric matrices, we omit elements below the diagonal.

The data were first analysed with $k = 2$. A proper solution was obtained. On testing the hypothesis H_k, the approximate χ^2 criterion U_k was found to be 27·53, with 19 degrees of freedom. On H_k the probability P of this value being exceeded is about 0·09. As a general rule, it is probably wise to increase the value of k until the value of χ^2 is below the 10 per cent significance level, i.e. until $P > 0{\cdot}10$. The analysis was therefore repeated with three factors.

Table 4.1 COVARIANCE MATRIX FOR DATA 1

1	2	3	4	5	6	7	8	9
1·000	0·523	0·395	0·471	0·346	0·426	0·576	0·434	0·639
	1·000	0·479	0·506	0·418	0·462	0·547	0·283	0·645
		1·000	0·355	0·270	0·254	0·452	0·219	0·504
			1·000	0·691	0·791	0·443	0·285	0·505
				1·000	0·679	0·383	0·149	0·409
					1·000	0·372	0·314	0·472
						1·000	0·385	0·680
							1·000	0·470
								1·000

For $k = 3$, the initial estimates of the residual variances, as given by (4.21), were as follows:

0·424, 0·414, 0·566, 0·250, 0·380, 0·267, 0·394, 0·595, 0·290. With these values the function f was found to be 0·086358. After two steepest descent iterations the value of f was reduced to 0·039075. The initial **E** matrix was then calculated and the Fletcher and Powell method put into operation. After 5 iterations the value of f was further reduced to 0·035017. This is the minimum value, since none of the derivations $\partial f/\partial\psi_i$ then exceeded 0·00001 in absolute magnitude. In *Table 4.2* we give the maximum likelihood estimates. The ith column of $\hat{\boldsymbol{\Lambda}}$ is denoted by $\hat{\lambda}_i$ and $\hat{\boldsymbol{\Psi}}$ is given in column form.

Table 4.2 MAXIMUM LIKELIHOOD SOLUTION FOR DATA 1

Variate	$\hat{\boldsymbol{\lambda}}_1$	$\hat{\boldsymbol{\lambda}}_2$	$\hat{\boldsymbol{\lambda}}_3$	$\hat{\boldsymbol{\Psi}}$
1	0·664	0·321	0·074	0·450
2	0·689	0·247	−0·193	0·427
3	0·493	0·302	−0·222	0·617
4	0·837	−0·292	−0·035	0·212
5	0·705	−0·315	−0·153	0·381
6	0·819	−0·377	0·105	0·177
7	0·661	0·396	−0·078	0·400
8	0·458	0·296	0·491	0·462
9	0·766	0·427	−0·012	0·231

For $k = 3$ the multiplying factor in (4.30) is 204·17, giving $U_k = 7{\cdot}149$. The number d_k of degrees of freedom, as found from (4.31) is 12. Thus the value of the test criterion is well below expectation and the hypothesis H_k is accepted.

The latent roots of $\hat{\mathbf{\Psi}}^{-1/2}\mathbf{S}\hat{\mathbf{\Psi}}^{-1/2}$ are:

15·968, 4·358, 1·848, 1·156, 1·119, 1·027, 0·926, 0·895, 0·877. The last six roots are all fairly close to unity and it can be verified that their average value is 1·000. If we subtract unity from each of the first three roots, we get a good idea of the relative importance of the three factors. The third factor, though relatively unimportant, has a fairly large loading in variate *8*.

The residual covariance matrix $(\mathbf{S}-\hat{\mathbf{\Lambda}}\hat{\mathbf{\Lambda}}')$, with the three factors eliminated, is as given in *Table 4.3*. The diagonal elements are those of $\hat{\mathbf{\Psi}}$. The non-diagonal elements are small and do not differ significantly from zero.

Table 4.3 RESIDUAL COVARIANCE MATRIX FOR DATA 1

1	*2*	*3*	*4*	*5*	*6*	*7*	*8*	*9*
0·450	0·000	−0·013	0·011	−0·010	−0·005	0·015	−0·001	−0·006
	0·427	0·022	−0·005	−0·019	0·011	−0·022	−0·011	0·010
		0·617	0·023	−0·016	−0·012	−0·011	0·013	−0·005
			0·212	0·003	−0·001	0·002	0·005	−0·011
				0·381	−0·001	0·029	−0·006	0·002
					0·177	−0·012	−0·001	0·007
						0·400	0·003	0·003
							0·462	−0·001
								0·231

The second set of data is that given by Maxwell (1961). It consists of 810 observations of 10 variates. Thus $n = 809$ and $p = 10$. For convenience we have changed the signs of variates *6–10* in order to make the covariances all positive. The standardised covariance matrix is then as given in *Table 4.4*.

Table 4.4 COVARIANCE MATRIX FOR DATA 2

1	*2*	*3*	*4*	*5*	*6*	*7*	*8*	*9*	*10*
1·000	0·345	0·594	0·404	0·579	0·280	0·449	0·188	0·303	0·200
	1·000	0·477	0·338	0·230	0·159	0·205	0·120	0·168	0·145
		1·000	0·498	0·505	0·251	0·377	0·186	0·273	0·154
			1·000	0·389	0·168	0·249	0·173	0·195	0·055
				1·000	0·151	0·285	0·129	0·159	0·079
					1·000	0·363	0·359	0·227	0·260
						1·000	0·448	0·439	0·511
							1·000	0·429	0·316
								1·000	0·301
									1·000

The data were first analysed with $k = 3$. The maximum likelihood solution was a proper one, but the test of H_k gave a χ^2 value of 78·545, which with 18 degrees of freedom is highly significant. The analysis was therefore repeated with four factors instead of three.

For $k = 4$ the calculations proceeded in the same way as before until the third iteration with the Fletcher and Powell method, when ψ_8 became equal to 0·005. It was then fixed at this value and the function f was minimised with respect to the other nine elements of $\mathbf{\Psi}$. This required another 5 iterations. The final value of f was 0·022849 and that of $\partial f/\partial \psi_8$ was 0·01013. All the other first-order derivatives were very close to zero. The maximum likelihood solution, an improper one, is shown in *Table 4.5*.

Table 4.5 IMPROPER SOLUTION FOR **DATA** 2

Variate	$\hat{\boldsymbol{\lambda}}_1$	$\hat{\boldsymbol{\lambda}}_2$	$\hat{\boldsymbol{\lambda}}_3$	$\hat{\boldsymbol{\lambda}}_4$	$\hat{\mathbf{\Psi}}$
1	0·198	0·750	−0·033	−0·108	0·385
2	0·126	0·467	−0·102	0·365	0·623
3	0·196	0·765	−0·165	0·217	0·301
4	0·179	0·524	−0·200	0·123	0·638
5	0·137	0·671	−0·249	−0·349	0·347
6	0·364	0·254	0·152	0·048	0·778
7	0·458	0·496	0·505	−0·052	0·286
8	0·997	−0·013	−0·005	0·000	0·005
9	0·435	0·275	0·207	0·053	0·690
10	0·322	0·226	0·495	0·029	0·600

The variate *8* was now eliminated and a proper solution obtained for the remaining nine variates. The initial values of the ψ_i were taken from *Table 4.5*. Only one further iteration with the Fletcher and Powell method was required. The final value of f was 0·022800. None of the derivatives exceeded 0·00002 in absolute value. The first three latent roots of $\hat{\mathbf{\Psi}}^{-1/2}\mathbf{S}\hat{\mathbf{\Psi}}^{-1/2}$ were 7·692, 2·757 and 1·794.

The hypothesis H_k was tested, with n, p and k all reduced by 1. Thus $n = 808$, $p = 9$ and $k = 3$. The multiplying factor in (4.30) is 802·17. Hence the test criterion U_k is 18·289 and has 12 degrees of freedom. It is not significant at the 10 per cent level, the value of P being 0·11. Hence H_k is accepted.

Since the variates are standardised, the first factor is taken to be identical with x_8. Hence the loadings of the variates on the first factor are simply the elements of the eighth column of the covariance matrix. The proper solution provides the residual variances and the loadings of all variates except x_8 on the second, third and fourth factors. The loadings of x_8 on these three factors are, of course, all zero. The final solution obtained by combining all these results is given in *Table 4.6*. As can be seen, it is very similar to the improper solution of *Table 4.5*.

Table 4.6 FINAL COMBINED SOLUTION FOR DATA 2

Variate	$\hat{\lambda}_1$	$\hat{\lambda}_2$	$\hat{\lambda}_3$	$\hat{\lambda}_4$	$\hat{\Psi}$
1	0·188	0·753	−0·035	−0·108	0·385
2	0·120	0·468	−0·103	0·365	0·623
3	0·186	0·767	−0·167	0·217	0·301
4	0·173	0·526	−0·200	0·124	0·638
5	0·129	0·672	−0·251	−0·349	0·347
6	0·359	0·259	0·154	0·048	0·778
7	0·448	0·504	0·507	−0·052	0·286
8	1·000	0·000	0·000	0·000	0·000
9	0·429	0·282	0·209	0·053	0·690
10	0·316	0·232	0·496	0·029	0·600

Since the final solution with 4 factors was accepted, there is no practical advantage to be gained by introducing a fifth factor. However, the data were in fact analysed with $k = 5$, and it is interesting to note that an improper solution again occurred. This time both ψ_6 and ψ_8 had values of 0·005. When x_6 and x_8 were eliminated the acceptability of the resulting proper solution was in doubt, as the corresponding value of P was now only 0·07.

4.7 ALTERNATIVE METHOD OF MINIMISING

The method of minimising $f_k(\mathbf{\Psi})$ described in section 4.3 has been found adequate for all data so far encountered. However, exact expressions for the second-order derivatives $\partial^2 f/\partial\psi_i\partial\psi_j$ have now been obtained by Clarke (1970), and these have enabled him to use the Newton–Raphson method (see Appendix II). Though the exact expressions are somewhat complicated and necessitate a greater amount of computation in each iteration, there is some evidence that considerably fewer iterations are usually required than with the procedure of section 4.3. Thus it seems likely that the Newton–Raphson method is computationally more efficient than that of Fletcher and Powell except in cases where p is fairly large. Some further details are given in section A2.7 of Appendix II.

Exercises

4.1. By use of the result (A1.9) of Appendix I, with a suitable choice of **A** and **B**, show that the determinant of the matrix

$$\mathbf{\Sigma} = \mathbf{\Lambda}\mathbf{\Lambda}' + \mathbf{\Psi}$$

is equal to

$$|\mathbf{\Psi}|\,|\mathbf{I} + \mathbf{\Lambda}'\mathbf{\Psi}^{-1}\mathbf{\Lambda}|.$$

4.2. The loadings given in *Table 6.2* of Chapter 6 represent a maximum likelihood solution, with $k = 2$, for the matrix **S** of *Table 6.1*. Find the residual covariance matrix $\mathbf{S} - \hat{\mathbf{\Lambda}}\hat{\mathbf{\Lambda}}'$. Verify that none of its non-diagonal elements exceeds 0·030 in absolute magnitude.

4.3. For the analysis considered in Exercise 4.2 the value of $f(\hat{\mathbf{\Psi}})$ is 0·010867. Use the criterion of (4.30) to test the hypothesis that two common factors are sufficient to account for the data. (The value of n is 219.)

4.4. Suppose that the ψ_i are known to be equal, but that the common value ψ is unknown. By putting $\mathbf{\Psi} = \psi\mathbf{I}$ and employing the methods of section 4.2, show that

$$\partial F/\partial\psi = \text{tr}\,[\mathbf{\Sigma}^{-1}(\mathbf{\Sigma} - \mathbf{S})\mathbf{\Sigma}^{-1}].$$

Using also the expression of (4.3) for $\partial F/\partial\mathbf{\Lambda}$, show that the maximum likelihood estimates of ψ and of $\mathbf{\Lambda}$ are given by

$$\hat{\psi} = [\text{tr}\,\mathbf{S} - (\gamma_1 + \ldots + \gamma_k)]/(p-k),$$

$$\mathbf{S}\hat{\mathbf{\Lambda}} = \hat{\mathbf{\Lambda}}\mathbf{\Gamma},$$

$$\hat{\mathbf{\Lambda}}'\hat{\mathbf{\Lambda}} = \mathbf{\Gamma} - \hat{\psi}\mathbf{I},$$

where $\mathbf{\Gamma}$ is a diagonal matrix of order k whose diagonal elements $\gamma_1, \ldots, \gamma_k$ are the k largest latent roots of **S**. Thus the columns $\hat{\boldsymbol{\lambda}}_1, \ldots, \hat{\boldsymbol{\lambda}}_k$ of $\hat{\mathbf{\Lambda}}$ are the corresponding latent vectors of **S** scaled in such a way that

$$\hat{\boldsymbol{\lambda}}_r'\hat{\boldsymbol{\lambda}}_r = \gamma_r - \hat{\psi}.$$

Chapter 5

SAMPLING FORMULAE FOR THE UNRESTRICTED MODEL

5.1 SAMPLING VARIANCES AND COVARIANCES OF ESTIMATES

In this chapter we find the asymptotic sampling variances and covariances of the maximum likelihood estimates of the elements of $\mathbf{\Lambda}$ and $\mathbf{\Psi}$. We assume that n is large and that the matrix $\mathbf{S}$ is unstandardised. The effect of standardisation will be considered in section 5.3. The notation of section 4.5 of Chapter 4 is used and all equations are correct to the same order of approximation as there. It is assumed that $\mathbf{\Lambda}$ and $\mathbf{\Psi}$ are uniquely defined and that all diagonal elements of $\mathbf{\Psi}$ are positive.

We begin by considering the maximum likelihood estimates $\hat{\psi}_1, \ldots, \hat{\psi}_p$. By standard estimation theory these have an asymptotic multivariate normal distribution with means given by $E(\hat{\psi}_i) = \psi_i$. The covariance matrix is the inverse of the 'information matrix', which has elements

$$E[(\partial \log L/\partial\psi_i)\,(\partial \log L/\partial\psi_j)] = -E(\partial^2 \log L/\partial\psi_i\,\partial\psi_j),$$

where L is evaluated with $\mathbf{\Lambda} = \hat{\mathbf{\Lambda}}_0$. Since $\log L$ is, apart from a function of the observations, the same as

$$-\tfrac{1}{2}nF(\hat{\mathbf{\Lambda}}_0, \mathbf{\Psi}) = -\tfrac{1}{2}nf(\mathbf{\Psi}),$$

the information matrix has elements

$$\tfrac{1}{2}nE(\partial^2 f/\partial\psi_i\partial\psi_j)$$

and, by (4.48) and (4.49), is therefore $\frac{1}{2}n\mathbf{G}$. The required covariance matrix is thus $(2/n)\mathbf{E}$, where $\mathbf{E} = \mathbf{G}^{-1}$, as before.

To evaluate $\mathbf{E}$ in practice we can either use the final $\mathbf{E}$ matrix obtained with the Fletcher and Powell method or, more accurately, estimate $\mathbf{\Phi}$ as

$$\hat{\mathbf{\Psi}}^{-1/2}(\mathbf{I} - \hat{\mathbf{\Omega}}\hat{\mathbf{\Omega}}')\hat{\mathbf{\Psi}}^{-1/2}$$

and then recalculate $\mathbf{G}$ by $g_{ij} = \phi_{ij}^2$ and $\mathbf{E}$ by inversion of $\mathbf{G}$.

For data 1, analysed in section 4.6 of Chapter 4, with $k = 3$, the **E** matrix estimated by the latter method is as given in *Table 5.1*.

Table 5.1 ESTIMATED **E** MATRIX FOR DATA 1

0·29	0·00	0·01	0·00	0·01	0·01	−0·01	−0·15	−0·02
	0·31	−0·02	−0·01	0·00	0·01	−0·01	−0·09	−0·02
		0·55	0·00	0·01	0·00	−0·01	−0·12	−0·01
			0·15	−0·02	−0·10	0·00	0·11	0·00
				0·27	0·00	0·00	−0·19	0·01
					0·25	−0·01	−0·25	0·01
						0·26	0·04	−0·03
							2·99	−0·10
								0·16

Let us for the present ignore the fact that **S** was standardised. Then the standard error of $\hat{\psi}_i$ is estimated as $\sqrt{(2e_{ii}/n)}$, where $n = 210$ and where e_{ij} denotes the element in the ith row and jth column of **E**. The largest standard error is that for $\hat{\psi}_8 = 0{\cdot}46$, which has the value 0·17. The others are much smaller. The covariances, which correspond to non-diagonal elements of **E**, are in general very small.

Before considering the maximum likelihood estimates of the λ_{ir} we require some preliminary results. Let us define the matrix **U** by

$$\mathbf{U} = \mathbf{S} - \mathbf{\Sigma}.$$

Then the $\frac{1}{2}p(p+1)$ distinct elements of $\sqrt{n}\mathbf{U}$ are known to have an asymptotic multivariate normal distribution with zero means and with variances and covariances given by

$$nE(u_{ij}u_{hl}) = \sigma_{ih}\sigma_{jl} + \sigma_{il}\sigma_{jh}. \tag{5.1}$$

A proof of this result may be found in various books. See, for example, that by Anderson (1958).

By use of (5.1) we establish a further result, namely that if $\boldsymbol{\alpha}$, $\boldsymbol{\beta}$, $\boldsymbol{\gamma}$, and $\boldsymbol{\delta}$ are any column vectors each having p elements, the covariance of $\boldsymbol{\alpha}'\mathbf{U}\boldsymbol{\beta}$ and $\boldsymbol{\gamma}'\mathbf{U}\boldsymbol{\delta}$, which are linear functions of the elements of **U**, is

$$(1/n)[(\boldsymbol{\alpha}'\mathbf{\Sigma}\boldsymbol{\gamma})(\boldsymbol{\beta}'\mathbf{\Sigma}\boldsymbol{\delta}) + (\boldsymbol{\alpha}'\mathbf{\Sigma}\boldsymbol{\delta})(\boldsymbol{\beta}'\mathbf{\Sigma}\boldsymbol{\gamma})]. \tag{5.2}$$

To prove this, let the ith elements of $\boldsymbol{\alpha}$, $\boldsymbol{\beta}$, $\boldsymbol{\gamma}$ and $\boldsymbol{\delta}$ be α_i, β_i, γ_i and δ_i respectively and let $\sum$ denote summation over all suffices. Then

n times the required covariance is given by

$$\begin{aligned} &nE[(\boldsymbol{\alpha}'\mathbf{U}\boldsymbol{\beta})(\boldsymbol{\gamma}'\mathbf{U}\boldsymbol{\delta})] \\ &= nE[\textstyle\sum(\alpha_i u_{ij}\beta_j)(\gamma_h u_{hl}\delta_l)] \\ &= n\textstyle\sum[\alpha_i\beta_j\gamma_h\delta_l E(u_{ij}u_{hl})] \\ &= \textstyle\sum[\alpha_i\beta_j\gamma_h\delta_l(\sigma_{ih}\sigma_{jl}+\sigma_{il}\sigma_{jh})] \\ &= \textstyle\sum[(\alpha_i\sigma_{ih}\gamma_h)(\beta_j\sigma_{jl}\delta_l)+(\alpha_i\sigma_{il}\delta_l)(\beta_j\sigma_{jh}\gamma_h)] \\ &= (\boldsymbol{\alpha}'\boldsymbol{\Sigma}\boldsymbol{\gamma})(\boldsymbol{\beta}'\boldsymbol{\Sigma}\boldsymbol{\delta})+(\boldsymbol{\alpha}'\boldsymbol{\Sigma}\boldsymbol{\delta})(\boldsymbol{\beta}'\boldsymbol{\Sigma}\boldsymbol{\gamma}). \end{aligned}$$

This establishes the result given in (5.2).

The result can be used in various forms. Suppose, for example, that $\boldsymbol{\alpha}$, $\boldsymbol{\beta}$, $\boldsymbol{\gamma}$ and $\boldsymbol{\delta}$ are columns i, j, h and l of the matrices $\mathbf{A}$, $\mathbf{B}$, $\mathbf{C}$ and $\mathbf{D}$ respectively. Then $(\boldsymbol{\alpha}'\mathbf{U}\boldsymbol{\beta})$ and $(\boldsymbol{\gamma}'\mathbf{U}\boldsymbol{\delta})$ become respectively $(\mathbf{A}'\mathbf{U}\mathbf{B})_{ij}$ and $(\mathbf{C}'\mathbf{U}\mathbf{D})_{hl}$, and their covariance is

$$(1/n)[(\mathbf{A}'\boldsymbol{\Sigma}\mathbf{C})_{ih}(\mathbf{B}'\boldsymbol{\Sigma}\mathbf{D})_{jl}+(\mathbf{A}'\boldsymbol{\Sigma}\mathbf{D})_{il}(\mathbf{B}'\boldsymbol{\Sigma}\mathbf{C})_{jh}]. \tag{5.3}$$

If we take $\mathbf{A}$ to be the unit matrix, we find that the covariance of $(\mathbf{UB})_{ij}$ and $(\mathbf{C}'\mathbf{UD})_{hl}$ is given by (5.3) with $(\mathbf{A}'\boldsymbol{\Sigma}\mathbf{C})_{ih}$ and $(\mathbf{A}'\boldsymbol{\Sigma}\mathbf{D})_{il}$ replaced by $(\boldsymbol{\Sigma}\mathbf{C})_{ih}$ and $(\boldsymbol{\Sigma}\mathbf{D})_{il}$ respectively. If both $\mathbf{A}$ and $\mathbf{C}$ are taken to be unit matrices, we find that the covariance of $(\mathbf{UB})_{ij}$ and $(\mathbf{UD})_{hl}$ is

$$(1/n)[\sigma_{ih}(\mathbf{B}'\boldsymbol{\Sigma}\mathbf{D})_{jl}+(\boldsymbol{\Sigma}\mathbf{D})_{il}(\boldsymbol{\Sigma}\mathbf{B})_{hj}]. \tag{5.4}$$

We consider first the errors of estimating $\boldsymbol{\Lambda}$ when $\boldsymbol{\Psi}$ is known. These are represented by the matrix $\mathbf{Z} = [z_{ir}]$, defined as in section 4.5 by

$$\mathbf{Z} = \hat{\boldsymbol{\Lambda}}_0-\boldsymbol{\Lambda}.$$

We express the elements of $\mathbf{Z}$ in terms of those of $\mathbf{U}$. To do this we rewrite equation (4.40), using (4.41), in the form

$$(\boldsymbol{\Lambda}\mathbf{Z}'+\mathbf{Z}\boldsymbol{\Lambda}'-\mathbf{U})\boldsymbol{\Psi}^{-1}\boldsymbol{\Lambda} = \mathbf{0}. \tag{5.5}$$

Since $\boldsymbol{\Psi}$ is assumed to be known, we may for convenience in carrying out the algebra take it to be the unit matrix. This represents merely a change in the metric of $\mathbf{x}$. We may then write

$$\boldsymbol{\Sigma} = \boldsymbol{\Lambda}\boldsymbol{\Lambda}'+\mathbf{I}, \tag{5.6}$$

$$\boldsymbol{\Theta} = \boldsymbol{\Lambda}'\boldsymbol{\Lambda}+\mathbf{I}, \tag{5.7}$$

$$\hat{\boldsymbol{\Theta}}_0 = \hat{\boldsymbol{\Lambda}}_0'\hat{\boldsymbol{\Lambda}}_0+\mathbf{I}, \tag{5.8}$$

and equation (5.5) becomes

$$\mathbf{Z}(\boldsymbol{\Theta}-\mathbf{I}) = \mathbf{U}\boldsymbol{\Lambda}-\boldsymbol{\Lambda}(\mathbf{Z}'\boldsymbol{\Lambda}). \tag{5.9}$$

By equating the elements in the ith row and rth column of the matrices on both sides we have

$$\begin{aligned} (\theta_r-1)z_{ir} &= (\mathbf{U}\boldsymbol{\Lambda})_{ir}-[\boldsymbol{\Lambda}(\mathbf{Z}'\boldsymbol{\Lambda})]_{ir} \\ &= (\mathbf{U}\boldsymbol{\Lambda})_{ir}-\sum_m [\lambda_{im}(\mathbf{Z}'\boldsymbol{\Lambda})_{mr}]. \end{aligned} \tag{5.10}$$

Let us suppose for the present that $k > 1$. From (5.7) and (5.8) we have

$$\begin{aligned}\hat{\mathbf{\Theta}}_0 - \mathbf{\Theta} &= \hat{\mathbf{\Lambda}}_0'\hat{\mathbf{\Lambda}}_0 - \mathbf{\Lambda}'\mathbf{\Lambda} \\ &= (\mathbf{\Lambda}'+\mathbf{Z}')(\mathbf{\Lambda}+\mathbf{Z}) - \mathbf{\Lambda}'\mathbf{\Lambda} \\ &= \mathbf{\Lambda}'\mathbf{Z}+\mathbf{Z}'\mathbf{\Lambda}.\end{aligned}$$

Since this matrix is diagonal, we have

$$(\mathbf{\Lambda}'\mathbf{Z})_{rs} = -(\mathbf{Z}'\mathbf{\Lambda})_{rs}, \quad \text{for} \quad r \neq s. \tag{5.11}$$

Now pre-multiply (5.9) by $\mathbf{\Lambda}'$ and use (5.7). Then with a rearrangement of terms we have

$$(\mathbf{\Lambda}'\mathbf{Z})(\mathbf{\Theta}-\mathbf{I})+(\mathbf{\Theta}-\mathbf{I})(\mathbf{Z}'\mathbf{\Lambda}) = \mathbf{\Lambda}'\mathbf{U}\mathbf{\Lambda}.$$

By equating the rth diagonal elements of the matrices on both sides of this we find that

$$2(\theta_r-1)(\mathbf{Z}'\mathbf{\Lambda})_{rr} = (\mathbf{\Lambda}'\mathbf{U}\mathbf{\Lambda})_{rr}. \tag{5.12}$$

For $r \neq s$, by equating the elements in the rth row and sth column on both sides, we have also

$$(\theta_s-1)(\mathbf{\Lambda}'\mathbf{Z})_{rs}+(\theta_r-1)(\mathbf{Z}'\mathbf{\Lambda})_{rs} = (\mathbf{\Lambda}'\mathbf{U}\mathbf{\Lambda})_{rs},$$

which by use of (5.11) simplifies to

$$(\theta_r-\theta_s)(\mathbf{Z}'\mathbf{\Lambda})_{rs} = (\mathbf{\Lambda}'\mathbf{U}\mathbf{\Lambda})_{rs}. \tag{5.13}$$

Equations (5.12) and (5.13) may be written respectively as

$$(\mathbf{Z}'\mathbf{\Lambda})_{rr} = \tfrac{1}{2}w_{rr}/(\theta_r-1)$$

and

$$(\mathbf{Z}'\mathbf{\Lambda})_{rs} = w_{rs}/(\theta_r-\theta_s) \qquad (r \neq s),$$

where

$$w_{rs} = w_{sr} = (\mathbf{\Lambda}'\mathbf{U}\mathbf{\Lambda})_{rs}. \tag{5.14}$$

Hence, substituting for the elements of $(\mathbf{Z}'\mathbf{\Lambda})$ in (5.10), we have

$$(\theta_r-1)z_{ir} = v_{ir}-\tfrac{1}{2}\lambda_{ir}w_{rr}/(\theta_r-1)+\sum_m{}' \,[\lambda_{im}w_{mr}/(\theta_r-\theta_m)], \tag{5.15}$$

where

$$v_{ir} = (\mathbf{U}\mathbf{\Lambda})_{ir} \tag{5.16}$$

and where $\sum_m{}'$ means that m is to be summed over all values from 1 to k except r. If $k = 1$, then r can take only the value 1 and the last term on the right-hand side of (5.15) disappears.

Since the variables v_{ir} and w_{rs} are linear functions of the elements of $\mathbf{U}$, so also are the variables z_{ir}. Hence the z_{ir} have an asymptotic multivariate normal distribution with zero means. Due to the restrictions on the elements of $\mathbf{Z}$, represented by (5.11), the covariance matrix is, however, singular.

To find the variances and covariances of the z_{ir} we require various covariances involving the variables v_{ir} and w_{ms}. These are obtained by use of (5.3) and (5.4). As an example consider the covariance of v_{ir} and v_{js}. This covariance, multiplied by n, is given by

$$nE(v_{ir}v_{js}) = nE[(\mathbf{U\Lambda})_{ir}(\mathbf{U\Lambda})_{js}]$$
$$= \sigma_{ij}(\mathbf{\Lambda'\Sigma\Lambda})_{rs} + (\mathbf{\Sigma\Lambda})_{is}(\mathbf{\Sigma\Lambda})_{jr}.$$

This result can be simplified since, from (5.6) and (5.7), we have

$$\mathbf{\Sigma\Lambda} = \mathbf{\Lambda\Theta},$$
$$\mathbf{\Lambda'\Sigma\Lambda} = \mathbf{\Theta}^2 - \mathbf{\Theta},$$

and hence

$$(\mathbf{\Sigma\Lambda})_{ir} = \theta_r\lambda_{ir},$$
$$(\mathbf{\Lambda'\Sigma\Lambda})_{rr} = \theta_r(\theta_r - 1),$$
$$(\mathbf{\Lambda'\Sigma\Lambda})_{rs} = 0 \qquad (r \neq s).$$

Other covariances are obtained in a similar manner. The results required are as follows:

$$nE(v_{ir}v_{jr}) = \theta_r[(\theta_r - 1)\sigma_{ij} + \theta_r\lambda_{ir}\lambda_{jr}],$$
$$nE(v_{ir}w_{rr}) = 2\theta_r^2(\theta_r - 1)\lambda_{ir},$$
$$nE(w_{rr}^2) = 2\theta_r^2(\theta_r - 1)^2,$$

and for $r \neq s$,

$$nE(v_{ir}v_{js}) = \theta_r\theta_s\lambda_{is}\lambda_{jr},$$
$$nE(v_{ir}w_{rs}) = \theta_r\theta_s(\theta_r - 1)\lambda_{is},$$
$$nE(w_{rs}^2) = \theta_r\theta_s(\theta_r - 1)(\theta_s - 1).$$

Any other choice of suffices gives rise to a zero covariance.

Use of the above results enables us to find the variances and covariances of the elements $\hat{\lambda}_{ir0}$ of $\hat{\mathbf{\Lambda}}_0$.

Let us write

$$a_{ir,js} = nE(z_{ir}z_{js}) = nE[(\hat{\lambda}_{ir0} - \lambda_{ir})(\hat{\lambda}_{js0} - \lambda_{js})].$$

Thus $a_{ir,js}$ is n times the covariance of $\hat{\lambda}_{ir0}$ and $\hat{\lambda}_{js0}$. Then, with a little algebra, we find from (5.15) that

$$a_{ir,jr} = \mu_r[\sigma_{ij} - \tfrac{1}{2}\mu_r\lambda_{ir}\lambda_{jr} + \sum_m{}' (\mu_m\gamma_{rm}\lambda_{im}\lambda_{jm})], \tag{5.17}$$

where

$$\mu_r = \theta_r/(\theta_r - 1)$$

and

$$\gamma_{rm} = [(\theta_r - 1)/(\theta_r - \theta_m)]^2 - 1.$$

We also have

$$a_{ir,js} = -[\theta_r\theta_s/(\theta_r - \theta_s)^2]\lambda_{is}\lambda_{jr} \qquad (r \neq s). \tag{5.18}$$

In order to generalise the foregoing results to cases where $\mathbf{\Psi}$ is not necessarily the unit matrix, we replace $\mathbf{\Lambda}$ by $\mathbf{\Psi}^{-1/2}\mathbf{\Lambda}$, $\mathbf{Z}$ by $\mathbf{\Psi}^{-1/2}\mathbf{Z}$, $\mathbf{\Sigma}$ by $\mathbf{\Psi}^{-1/2}\mathbf{\Sigma}\mathbf{\Psi}^{-1/2}$ and $\mathbf{U}$ by $\mathbf{\Psi}^{-1/2}\mathbf{U}\mathbf{\Psi}^{-1/2}$. This involves corresponding adjustments in the elements of these matrices and in the various expressions that depend upon them. When all these adjustments are made it is found that equation (5.15) is correct as it stands provided that v_{ir} and w_{rs} are redefined by

$$v_{ir} = (\mathbf{U}\mathbf{\Psi}^{-1}\mathbf{\Lambda})_{ir} \tag{5.19}$$

and

$$w_{rs} = (\mathbf{\Lambda}'\mathbf{\Psi}^{-1}\mathbf{U}\mathbf{\Psi}^{-1}\mathbf{\Lambda})_{rs}. \tag{5.20}$$

The θ_r are the elements of $\mathbf{\Theta}$ as defined in section 4.5. Equations (5.17) and (5.18) remain unaltered.

If $k = 1$, the summation term in (5.17) disappears and r takes only the value 1. Equation (5.18) is then not applicable.

The above formulae were derived by Lawley (1953). Their usefulness is impaired by the fact that they relate to maximum likelihood estimates of the λ_{ir} when $\mathbf{\Psi}$ is known. We may regard $a_{ir,js}$ as being n times the partial covariance of $\hat{\lambda}_{ir}$ and $\hat{\lambda}_{js}$ for given $\mathbf{\Psi}$. To find the total covariance it is necessary to add the component arising from the linear regression of $\hat{\lambda}_{ir}$ and $\hat{\lambda}_{js}$ on the elements of $\hat{\mathbf{\Psi}}$. A method of doing this was given in a later paper (Lawley, 1967).

Let us regard the right-hand side of (5.15) divided by $(\theta_r - 1)$ as a function of the elements of $\mathbf{\Psi}$ and $\mathbf{S}$ and denote it by $g_{ir}(\mathbf{\Psi}, \mathbf{S})$. We include $\mathbf{S}$ to indicate that g_{ir} is a random variable. Equation (5.15) may then be written as

$$\hat{\lambda}_{ir0} - \lambda_{ir} = g_{ir}(\mathbf{\Psi}, \mathbf{S}). \tag{5.21}$$

If $\mathbf{\Psi}$ is replaced by $\hat{\mathbf{\Psi}}$, this means that maximum likelihood estimation is with respect to both $\mathbf{\Psi}$ and $\mathbf{\Lambda}$. Hence $\hat{\lambda}_{ir0}$ is replaced by $\hat{\lambda}_{ir}$ and we have

$$\hat{\lambda}_{ir} - \lambda_{ir} = g_{ir}(\hat{\mathbf{\Psi}}, \mathbf{S}). \tag{5.22}$$

By subtracting (5.21) from (5.22) and using the fact that the elements of $(\hat{\mathbf{\Psi}} - \mathbf{\Psi})$ are small in probability we have, to the usual order of approximation,

$$\begin{aligned} \hat{\lambda}_{ir} - \hat{\lambda}_{ir0} &= g_{ir}(\hat{\mathbf{\Psi}}, \mathbf{S}) - g_{ir}(\mathbf{\Psi}, \mathbf{S}) \\ &= \sum_j [(\partial g_{ir}/\partial \psi_j)(\hat{\psi}_j - \psi_j)]. \end{aligned}$$

Thus we may write

$$\begin{aligned} \hat{\lambda}_{ir} - \lambda_{ir} &= (\hat{\lambda}_{ir} - \hat{\lambda}_{ir0}) + (\hat{\lambda}_{ir0} - \lambda_{ir}) \\ &= \sum_j [b_{j,ir}(\hat{\psi}_j - \psi_j)] + z_{ir}, \end{aligned} \tag{5.23}$$

where $b_{j,ir}$ is the limit in probability of $\partial g_{ir}/\partial \psi_j$, as $n \to \infty$.

In the above equation $(\hat{\lambda}_{ir}-\lambda_{ir})$ is divided into two components. The first component $(\hat{\lambda}_{ir}-\hat{\lambda}_{ir0})$ represents the effect of the linear regression of $\hat{\lambda}_{ir}$ on $\hat{\psi}_1, \ldots, \hat{\psi}_p$. The regression coefficient of $\hat{\lambda}_{ir}$ on $\hat{\psi}_j$ is $b_{j,ir}$. The second component, $z_{ir} = \hat{\lambda}_{ir0}-\lambda_{ir}$, is independent of this regression. Asymptotically, the elements of $\mathbf{Z}$ are distributed independently of the elements of $(\hat{\mathbf{\Lambda}}-\hat{\mathbf{\Lambda}}_0)$. Let $\mathbf{b}_{ir}$ be the column vector whose elements are $b_{1,ir}, \ldots, b_{p,ir}$. Then, since the covariance matrix of $\hat{\psi}_1, \ldots, \hat{\psi}_p$ is $(2/n)\mathbf{E}$, the covariance of $(\hat{\lambda}_{ir}-\hat{\lambda}_{ir0})$ and $(\hat{\lambda}_{js}-\hat{\lambda}_{js0})$ is, by standard regression theory,

$$\mathbf{b}'_{ir}[(2/n)\mathbf{E}]\mathbf{b}_{js} = (2/n)\mathbf{b}'_{ir}\mathbf{E}\mathbf{b}_{js}. \tag{5.24}$$

Hence the covariance of $\hat{\lambda}_{ir}$ and $\hat{\lambda}_{js}$ is

$$(1/n)\,(a_{ir,js}+2\mathbf{b}'_{ir}\mathbf{E}\mathbf{b}_{js}). \tag{5.25}$$

For the variance of $\hat{\lambda}_{ir}$ we put $i = j$ and $r = s$.

To evaluate $b_{j,ir}$ we use the fact that the variables v_{ir} and w_{rs} are small in probability. Since g_{ir} is the right-hand side of (5.15) divided by (θ_r-1), we have to the usual order of approximation

$$\begin{aligned}(\theta_r-1)b_{j,ir} &= \partial v_{ir}/\partial\psi_j-\tfrac{1}{2}\lambda_{ir}(\theta_r-1)^{-1}\partial w_{rr}/\partial\psi_j\\ &\quad+\sum_m{}' [\lambda_{im}(\theta_r-\theta_m)^{-1}\partial w_{rm}/\partial\psi_j],\end{aligned} \tag{5.26}$$

where v_{ir} and w_{rs} are as given by (5.19) and (5.20). Since the elements of $\mathbf{U}$ are small in probability, we have also

$$\partial v_{ir}/\partial\psi_j = [(\partial\mathbf{U}/\partial\psi_j)\mathbf{\Psi}^{-1}\mathbf{\Lambda}]_{ir}$$

and

$$\partial w_{rs}/\partial\psi_j = [\mathbf{\Lambda}'\mathbf{\Psi}^{-1}(\partial\mathbf{U}/\partial\psi_j)\mathbf{\Psi}^{-1}\mathbf{\Lambda}]_{rs}.$$

Now $\mathbf{U}$ is defined by

$$\mathbf{U} = \mathbf{S}-\mathbf{\Lambda}\mathbf{\Lambda}'-\mathbf{\Psi}.$$

Hence all elements of the matrix $\partial\mathbf{U}/\partial\psi_j$ are zero except the jth diagonal element, which is -1. It follows that, approximately,

$$\partial v_{ir}/\partial\psi_j = -\delta_{ij}\lambda_{jr}\psi_j^{-1}$$

and

$$\partial w_{rs}/\partial\psi_j = -\lambda_{jr}\lambda_{js}\psi_j{}^{-2},$$

where $\delta_{ij} = 0$ if $i \neq j$ and $\delta_{ii} = 1$. Substituting these results in (5.26) we have

$$\begin{aligned}b_{j,ir} &= -\lambda_{jr}(\theta_r-1)^{-1}\psi_j^{-2}[\delta_{ij}\psi_j-\tfrac{1}{2}\lambda_{ir}\lambda_{jr}/(\theta_r-1)\\ &\quad+\sum_m{}' \lambda_{im}\lambda_{jm}/(\theta_r-\theta_m)].\end{aligned} \tag{5.27}$$

An alternative expression is

$$b_{j,\,ir} = -[\psi_i/(\theta_r-1)]^{1/2}\omega_{jr}\psi_j^{-1}[\delta_{ij}-\tfrac{1}{2}\omega_{ir}\omega_{jr} + \sum_m{}' \omega_{im}\omega_{jm}(\theta_m-1)/(\theta_r-\theta_m)]. \tag{5.28}$$

The summation terms on the right-hand sides of (5.27) and (5.28) disappear when $k = 1$.

Let us arrange the pk elements of $\mathbf{\Lambda}$ so that they form a vector

$$\{\lambda_{11}\ldots\lambda_{p1}\lambda_{12}\ldots\lambda_{p2}\ldots\lambda_{1k}\ldots\lambda_{pk}\}.$$

Thus λ_{ir} is element number $[p(r-1)+i]$ in this vector. Let the elements of other $p\times k$ matrices be arranged similarly. Then the covariance matrix of the variables z_{ir} is $(1/n)\mathbf{A}$, where $\mathbf{A}$ is the matrix of which $a_{ir,js}$ is the element in row number $[p(r-1)+i]$ and column number $[p(s-1)+j]$. It is symmetric, of order pk and positive semi-definite. The regression coefficients $b_{j,\,ir}$ form a $p\times pk$ matrix $\mathbf{B}$ of which $\mathbf{b}_{ir}$ is column number $[p(r-1)+i]$. With this notation the covariance matrix of the $\hat{\lambda}_{ir}$ is

$$(1/n)(\mathbf{A}+2\mathbf{B}'\mathbf{E}\mathbf{B}). \tag{5.29}$$

From (5.23) we have

$$\begin{aligned} E[(\hat{\lambda}_{ir}-\lambda_{ir})(\hat{\psi}_j-\psi_j)] &= E\sum_h b_{h,\,ir}(\hat{\psi}_h-\psi_h)(\hat{\psi}_j-\psi_j) \\ &= (2/n)\sum_h (e_{jh}b_{h,\,ir}) \\ &= (2/n)(\mathbf{EB})_{j,\,ir}. \end{aligned} \tag{5.30}$$

Thus the covariance of $\hat{\lambda}_{ir}$ and $\hat{\psi}_j$ is the element in the jth row and in column number $[p(r-1)+i]$ of the matrix $(2/n)\mathbf{EB}$.

The formulae obtained above are complicated in form and it seems unlikely that they will often be used in practice. However, for moderate values of p and k, their evaluation by computer presents no great difficulty and the results may sometimes be of interest. Except in artificial sampling experiments the true values of the parameters would be unknown and would be replaced by their maximum likelihood estimates.

A further complication arises when $k > 1$ and when the factors are rotated or transformed. The matrix of loadings on the new factors is then $\mathbf{\Lambda M}$, where $\mathbf{M}$ is the transformation matrix. If $\mathbf{M}$ is treated as a known matrix, it is easy to compute the sampling variances and covariances of the new loadings, since these are merely linear functions of the original loadings in which the coefficients are elements of $\mathbf{M}$. This is not strictly justifiable as in practice $\mathbf{M}$ is usually derived from the data. It would be almost impossible to take sampling errors in the elements of $\mathbf{M}$ into account. The only course is therefore to ignore them in the hope that they are relatively small.

5.2 THE ONE-FACTOR CASE

It is worth considering in more detail the case where $k = 1$, as the sampling formulae are then comparatively simple.

We denote the loading of x_i on the one common factor by λ_i, the suffix r being omitted. Similarly we write $\hat{\lambda}_i$ and $\hat{\lambda}_{i0}$ for the corresponding estimates. The matrix $\mathbf{\Omega}$ becomes a column vector $\boldsymbol{\omega}$ with elements $\omega_1, \ldots, \omega_p$ satisfying

$$\omega_i = (\theta-1)^{-1/2}\lambda_i\psi_i^{-1/2}, \tag{5.31}$$

where

$$\theta = 1+\sum_j (\lambda_j^2/\psi_j).$$

The matrix $\mathbf{\Phi}$ is given by

$$\mathbf{\Phi} = \mathbf{\Psi}^{-1/2}(\mathbf{I}-\boldsymbol{\omega}\boldsymbol{\omega}')\mathbf{\Psi}^{-1/2},$$

and thus

$$\phi_{ij} = (\delta_{ij}-\omega_i\omega_j)\,(\psi_i\psi_j)^{-1/2}.$$

We have also

$$g_{ij} = \phi_{ij}^2 = [\delta_{ij}(1-2\omega_i^2)+\omega_i^2\omega_j^2](\psi_i\psi_j)^{-1}.$$

Thus $\mathbf{G} = [g_{ij}]$ is given by

$$\mathbf{G} = \mathbf{\Psi}^{-1}(\mathbf{D}+\boldsymbol{\alpha}\boldsymbol{\alpha}')\mathbf{\Psi}^{-1},$$

where $\mathbf{D}$ is a diagonal matrix whose ith diagonal element is $(1-2\omega_i^2)$ and where $\boldsymbol{\alpha}$ is a column vector whose ith element is ω_i^2.

Let us assume for the present that $\omega_i^2 \neq \frac{1}{2}$ for all values of i. Then $\mathbf{D}$ has no zero diagonal elements and $\mathbf{E}$ is given by

$$\mathbf{E} = \mathbf{G}^{-1} = \mathbf{\Psi}[\mathbf{D}^{-1}-(1/\mu)\boldsymbol{\gamma}\boldsymbol{\gamma}']\mathbf{\Psi}, \tag{5.32}$$

where

$$\boldsymbol{\gamma} = \mathbf{D}^{-1}\boldsymbol{\alpha}$$

and

$$\mu = 1+\boldsymbol{\alpha}'\mathbf{D}^{-1}\boldsymbol{\alpha} = 1+\sum_j [\omega_j^4/(1-2\omega_j^2)].$$

Hence for the elements of $\mathbf{E}$ we have

$$e_{ij} = [\delta_{ij}\beta_i-(1/\mu)\gamma_i\gamma_j]\psi_i\psi_j, \tag{5.33}$$

where

$$\beta_i = 1/(1-2\omega_i^2),$$
$$\gamma_i = \omega_i^2/(1-2\omega_i^2).$$

The variances and covariances of $\hat{\psi}_1, \ldots, \hat{\psi}_p$ are then given by

$$nE[(\hat{\psi}_i-\psi_i)\,(\hat{\psi}_j-\psi_j)] = 2e_{ij}. \tag{5.34}$$

Equations (5.28), (5.30) and (5.17) are now used, with appropriate modifications. From (5.28) we find that the regression coefficient of $\hat{\lambda}_i$ on $\hat{\psi}_j$ is

$$\begin{aligned} b_{ji} &= -(\theta-1)^{-1/2}\omega_i\psi_i^{1/2}\psi_j^{-1}(\delta_{ij}-\tfrac{1}{2}\omega_j^2) \\ &= -(\theta-1)^{-1}\lambda_i\psi_j^{-1}(\delta_{ij}-\tfrac{1}{2}\omega_j^2). \end{aligned} \tag{5.35}$$

The coefficient b_{ji} is the element in the jth row and ith column of the $p \times p$ matrix $\mathbf{B}$.

The covariance of $\hat{\lambda}_i$ and $\hat{\psi}_j$ is given by (5.30) as $(2/n)(\mathbf{EB})_{ji}$. This may be evaluated by use of (5.33) and (5.35). With a little algebra it is found that

$$\begin{aligned} &nE(\hat{\lambda}_i-\lambda_i)(\hat{\psi}_j-\psi_j) \\ &\quad = 2(\mathbf{EB})_{ji} \\ &\quad = 2\sum_h (e_{jh}b_{hi}) \\ &\quad = (\theta-1)^{-1}\lambda_i\beta_i\psi_j[-2\delta_{ij}+(1/\mu)\gamma_j]. \end{aligned} \tag{5.36}$$

The covariance of $(\hat{\lambda}_i-\hat{\lambda}_{i0})$ and $(\hat{\lambda}_j-\hat{\lambda}_{j0})$ is $(2/n)(\mathbf{B'EB})_{ij}$. This may be evaluated by use of (5.35), (5.36) and (5.31). We have

$$\begin{aligned} &nE[(\hat{\lambda}_i-\hat{\lambda}_{i0})(\hat{\lambda}_j-\hat{\lambda}_{j0})] \\ &\quad = 2(\mathbf{B'EB})_{ij} \\ &\quad = 2\sum_h [b_{hi}(\mathbf{EB})_{hj}] \\ &\quad = 2(\theta-1)^{-1}\delta_{ij}\gamma_i\psi_i+\tfrac{1}{2}(\theta-1)^{-2}\lambda_i\lambda_j[1-(1/\mu)\beta_i\beta_j]. \end{aligned} \tag{5.37}$$

From (5.17) the covariance of $\hat{\lambda}_{i0}$ and $\hat{\lambda}_{j\theta}$ is a_{ij}/n, where

$$\begin{aligned} a_{ij} &= \theta(\theta-1)^{-1}[\lambda_i\lambda_j+\delta_{ij}\psi_i)-\tfrac{1}{2}\theta(\theta-1)^{-1}\lambda_i\lambda_j] \\ &= \theta(\theta-1)^{-1}\delta_{ij}\psi_i+\tfrac{1}{2}\lambda_i\lambda_j[1-(\theta-1)^{-2}]. \end{aligned} \tag{5.38}$$

The variances and covariances of $\hat{\lambda}_1, \ldots, \hat{\lambda}_p$ are now obtained by adding (5.37) and (5.38). With a slight rearrangement of terms we have

$$\begin{aligned} &nE(\hat{\lambda}_i-\lambda_i)(\hat{\lambda}_j-\lambda_j) \\ &\quad = \delta_{ij}\psi_i[1+(\theta-1)^{-1}\beta_i]+\tfrac{1}{2}\lambda_i\lambda_j[1-(1/\mu)(\theta-1)^{-2}\beta_i\beta_j]. \end{aligned} \tag{5.39}$$

The formulae of (5.33), (5.36) and (5.39) are remarkably simple. They could easily be evaluated with the use of a desk calculating machine.

We must now discuss what happens when one of the quantities $\omega_1^2, \ldots, \omega_p^2$ is equal to $\frac{1}{2}$. The possibility of two of them having the value $\frac{1}{2}$ can be ruled out. For suppose that $\omega_1^2 = \omega_2^2 = \frac{1}{2}$. Then, since $\omega_1^2+\ldots+\omega_p^2 = 1$, we must have $\omega_3 = \ldots = \omega_p = 0$. Hence λ_1 and λ_2 are the only non-zero loadings. By section 2.3 of Chapter

2 the parameters are then not uniquely defined, which is contrary to supposition.

Let us suppose that $\omega_1^2 = \frac{1}{2}$. Then $\mathbf{E} = \mathbf{G}^{-1}$ is no longer given by (5.32), since the first diagonal element of $\mathbf{D}$ is zero. In this case let $\mathbf{D}_1$ be the diagonal matrix of order $p-1$ whose diagonal elements are $(1-2\omega_2^2), \ldots, (1-2\omega_p^2)$ and let $\boldsymbol{\alpha}_1$ be the column vector with elements $\omega_2^2, \ldots, \omega_p^2$. Then $\mathbf{\Psi G \Psi}$ is given in partitioned form by

$$\mathbf{\Psi G \Psi} = \begin{bmatrix} 1/4 & \frac{1}{2}\boldsymbol{\alpha}_1' \\ \frac{1}{2}\boldsymbol{\alpha}_1 & \mathbf{D}_1 + \boldsymbol{\alpha}_1\boldsymbol{\alpha}_1' \end{bmatrix}.$$

For its inverse we have

$$\mathbf{\Psi}^{-1}\mathbf{E}\mathbf{\Psi}^{-1} = \begin{bmatrix} 4\mu_1 & -2\boldsymbol{\gamma}_1' \\ -2\boldsymbol{\gamma}_1 & \mathbf{D}_1^{-1} \end{bmatrix},$$

where

$$\boldsymbol{\gamma}_1 = \mathbf{D}_1^{-1}\boldsymbol{\alpha}_1$$

and

$$\mu_1 = 1 + \boldsymbol{\alpha}_1'\mathbf{D}_1^{-1}\boldsymbol{\alpha}_1 = 1 + \sum_{j=2}^{p} \omega_j^4/(1-2\omega_j^2).$$

The above result enables us to write down the variances and covariances of $\hat{\psi}_1, \ldots, \hat{\psi}_p$. For $i \neq 1$, $j \neq 1$ and $i \neq j$ we have

$$\begin{aligned} n \operatorname{var}(\hat{\psi}_1) &= 2e_{11} = 8\mu_1\psi_1^2, \\ n \operatorname{var}(\hat{\psi}_i) &= 2e_{ii} = 2\beta_i\psi_i^2, \\ n \operatorname{cov}(\hat{\psi}_1, \hat{\psi}_i) &= 2e_{1i} = -4\gamma_i\psi_1\psi_i, \\ n \operatorname{cov}(\hat{\psi}_i, \psi_j) &= 0. \end{aligned}$$

The other variances and covariances may most easily be obtained by finding the limits of the expressions in (5.36) and (5.39) when $\omega_1^2 \to \frac{1}{2}$. With i and j as before we have

$$\begin{aligned} n \operatorname{cov}(\hat{\lambda}_1, \hat{\psi}_1) &= -2(\theta-1)^{-1}(4\mu_1 - 1)\lambda_1\psi_1, \\ n \operatorname{cov}(\hat{\lambda}_1, \hat{\psi}_i) &= 4(\theta-1)^{-1}\gamma_i\lambda_1\psi_i, \\ n \operatorname{cov}(\hat{\lambda}_i, \psi_1) &= 2(\theta-1)^{-1}\beta_i\lambda_i\psi_1, \\ n \operatorname{cov}(\hat{\lambda}_i, \hat{\psi}_i) &= -2\beta_i\lambda_i\psi_i, \\ n \operatorname{cov}(\hat{\lambda}_i, \hat{\psi}_j) &= 0. \end{aligned}$$

We have also

$$\begin{aligned} n \operatorname{var}(\hat{\lambda}_1) &= [1 + (\theta-1)^{-1}(4\mu_1 - 1)]\psi_1 + \tfrac{1}{2}\lambda_1^2, \\ n \operatorname{var}(\hat{\lambda}_i) &= [1 + (\theta-1)^{-1}\beta_i]\psi_i + \tfrac{1}{2}\lambda_i^2, \\ n \operatorname{cov}(\hat{\lambda}_1, \hat{\lambda}_i) &= \tfrac{1}{2}[1 - 4(\theta-1)^{-2}\beta_i]\lambda_1\lambda_i, \\ n \operatorname{cov}(\hat{\lambda}_i, \hat{\lambda}_j) &= \tfrac{1}{2}\lambda_i\lambda_j. \end{aligned}$$

In practice these formulae would be used if ω_1^2 were very near to $\frac{1}{2}$.

It is of course possible for one of the quantities $\omega_1^2, \ldots, \omega_p^2$ to exceed $\frac{1}{2}$. Suppose that $\frac{1}{2} < \omega_1^2 < 1$. Then μ and $(1-2\omega_1^2)$, the first element of **D**, are both negative. However **G** and **E** are still positive definite and equations (5.32)–(5.39) remain valid.

5.3 THE EFFECT OF STANDARDISATION

In the two previous sections we assumed that $\hat{\mathbf{\Lambda}}$ and $\hat{\mathbf{\Psi}}$ were obtained from the analysis of the unstandardised sample covariance matrix **S**. However, in actual practice it is usual to standardise all variates, so that **S** is replaced by the sample correlation matrix having unit diagonal elements. The analysis of the correlation matrix results in standardised loadings $\hat{\lambda}_{ir}^*$ and standardised residual variances $\hat{\psi}_i^*$ satisfying

$$\hat{\lambda}_{ir}^* = \hat{\lambda}_{ir}/\sqrt{s_{ii}}$$

and

$$\hat{\psi}_i^* = \hat{\psi}_i/s_{ii}.$$

We shall show how the sampling variances and covariances of these standardised estimates may be found. For this purpose we require the covariance of $\hat{\lambda}_{ir}$ and s_{jj} and the covariance of $\hat{\psi}_i$ and s_{jj}.

The first-order derivatives of the function $f = f_k(\mathbf{\Psi})$ of Chapter 4 vanish when $\hat{\mathbf{\Psi}}$ is substituted for $\mathbf{\Psi}$. Hence, using (4.48) and (4.49), we have to the usual order of approximation

$$\begin{aligned}\partial f/\partial\psi_i &= \sum_j [(\partial^2 f/\partial\psi_i\partial\psi_j)(\psi_j-\hat{\psi}_j)] \\ &= -\sum_j [g_{ij}(\hat{\psi}_j-\psi_j)].\end{aligned}$$

Since $\mathbf{E} = [e_{ij}]$ is the inverse of $\mathbf{G} = [g_{ij}]$, we find, using (4.47), that

$$\begin{aligned}\hat{\psi}_i-\psi_i &= -\sum_h e_{ih}(\partial f/\partial\psi_h) \\ &= -\sum_h e_{ih}[\mathbf{\Phi}(\mathbf{\Sigma}-\mathbf{S})\mathbf{\Phi}]_{hh} \\ &= \sum_h e_{ih}(\mathbf{\Phi U \Phi})_{hh},\end{aligned}$$

where $\mathbf{U} = \mathbf{S}-\mathbf{\Sigma}$, as before, and where $\mathbf{\Phi}$ is given by (4.43). By use of the methods of section 5.1 we now have

$$\begin{aligned} n\,\mathrm{cov}\,(\hat{\psi}_i, s_{jj}) &= nE[(\hat{\psi}_i - \psi_i)u_{jj}] \\ &= n\sum_h e_{ih}E[(\mathbf{\Phi U \Phi})_{hh}u_{jj}] \\ &= 2\sum_h e_{ih}[(\mathbf{\Phi\Sigma})_{hj}]^2 \\ &= 2\sum_h e_{ih}[(\mathbf{\Phi\Psi})_{hj}]^2 \\ &= 2\psi_j^2 \sum_h (e_{ih}\phi_{hj}^2) \\ &= 2\psi_j^2 \sum_h (e_{ih}g_{hj}) \\ &= 2\delta_{ij}\psi_j^2. \end{aligned} \tag{5.40}$$

For the covariance of $\hat{\lambda}_{ir}$ and s_{jj} we have

$$E[(\hat{\lambda}_{ir} - \lambda_{ir})u_{jj}] = E[(\hat{\lambda}_{ir0} - \lambda_{ir})u_{jj}] + E[(\hat{\lambda}_{ir} - \hat{\lambda}_{ir0})u_{jj}]. \tag{5.41}$$

To evaluate the first term on the right-hand side of this we use the expression of (5.15) for $z_{ir} = \hat{\lambda}_{ir0} - \lambda_{ir}$ with v_{ir} and w_{rs} as given by (5.19) and (5.20). After some algebra, which we omit, it is found that

$$\begin{aligned} &nE[(\hat{\lambda}_{ir0} - \lambda_{ir})u_{jj}] \\ &\quad = \theta_r(\theta_r - 1)^{-1}\lambda_{jr}[2\sigma_{ij} - \theta_r(\theta_r - 1)^{-1}\lambda_{ir}\lambda_{jr} \\ &\qquad + 2\sum_s{}' \theta_s(\theta_r - \theta_s)^{-1}\lambda_{is}\lambda_{js}], \end{aligned} \tag{5.42}$$

where $\sum_s{}'$ means, as usual, that s is summed over all values from 1 to k except r.

Alternatively we may use the equation

$$s_{jj} = \sum_m (\hat{\lambda}_{jm}^2) + \hat{\psi}_j,$$

which follows from (4.19) on substituting $\hat{\mathbf{\Psi}}$ for $\mathbf{\Psi}$ and $\hat{\mathbf{\Lambda}}$ for $\mathbf{\Lambda}_0$. Since also

$$\sigma_{jj} = \sum_m (\lambda_{jm}^2) + \psi_j,$$

we have, approximately,

$$\begin{aligned} u_{jj} &= s_{jj} - \sigma_{jj} \\ &= 2\sum_m [\lambda_{jm}(\hat{\lambda}_{jm} - \lambda_{jm})] + (\hat{\psi}_j - \psi_j) \\ &= 2\sum_m (\lambda_{jm}z_{jm}) + 2\sum_m [\lambda_{jm}(\hat{\lambda}_{jm} - \hat{\lambda}_{jm0})] + (\hat{\psi}_j - \psi_j). \end{aligned}$$

Hence, as z_{ir} is independent of both $(\hat{\lambda}_{jm}-\hat{\lambda}_{jm0})$ and $\hat{\psi}_j$, for all j and m, we have

$$\begin{aligned} nE[(\hat{\lambda}_{ir0}-\lambda_{ir})u_{jj}] &= 2n\sum_m [\lambda_{jm}E(z_{ir}z_{jm})] \\ &= 2\sum_m (\lambda_{jm}a_{ir,\,jm}), \end{aligned} \tag{5.43}$$

where $a_{ir,\,jm}$ is as given by (5.17) or (5.18). In practice this expression is probably more useful than that of (5.42) if all elements of **A** have been found.

To evaluate the second term on the right-hand side of (5.41) we use the equation

$$\hat{\lambda}_{ir}-\hat{\lambda}_{ir0} = \sum_h [b_{h,\,ir}(\hat{\psi}_h-\psi_h)].$$

Hence, using (5.40), we have

$$\begin{aligned} nE[(\hat{\lambda}_{ir}-\hat{\lambda}_{ir0})u_{jj}] &= n\sum_h b_{h,\,ir}E[(\hat{\psi}_h-\psi_h)u_{jj}] \\ &= \sum_h b_{h,\,ir}(2\delta_{hj}\psi_j^2) \\ &= 2\psi_j^2 b_{j,\,ir}. \end{aligned} \tag{5.44}$$

On adding (5.43) and (5.44) we have

$$\begin{aligned} n\operatorname{cov}(\hat{\lambda}_{ir}, s_{jj}) &= nE[(\hat{\lambda}_{ir}-\lambda_{ir})u_{jj}] \\ &= 2\sum_m (\lambda_{jm}a_{ir,\,jm})+2\psi_j^2 b_{j,\,ir}. \end{aligned} \tag{5.45}$$

The second term on the right-hand side may be written explicitly, using (5.27), as

$$\begin{aligned} 2\psi_j^2 b_{j,\,ir} = {} & (\theta_r-1)^{-1}\lambda_{jr}[-2\delta_{ij}\psi_j+(\theta_r-1)^{-1}\lambda_{ir}\lambda_{jr} \\ & -2\sum_s{}' (\theta_r-\theta_s)^{-1}\lambda_{is}\lambda_{js}]. \end{aligned} \tag{5.46}$$

Hence, adding (5.42) and (5.46), we find, with some rearrangement of terms,

$$\begin{aligned} n\operatorname{cov}(\hat{\lambda}_{ir}, s_{jj}) = {} & (\theta_r-1)^{-1}\lambda_{jr}[2\theta_r\sigma_{ij}-2\delta_{ij}\psi_i-(\theta_r+1)\lambda_{ir}\lambda_{jr} \\ & +2\sum_s{}' (\theta_r\theta_s-1)(\theta_r-\theta_s)^{-1}\lambda_{is}\lambda_{js}]. \end{aligned} \tag{5.47}$$

The expression in (5.45) is likely to be rather more convenient if the elements of the matrices **A** and **B** have already been found. As usual, the summation terms in (5.42), (5.46) and (5.47) disappear if $k = 1$.

Having obtained the above preliminary results, we now approximate the standardised estimates by

$$\begin{aligned} \hat{\lambda}_{ir}^* &= \hat{\lambda}_{ir}/\sqrt{s_{ii}} \\ &= \hat{\lambda}_{ir}/\sqrt{\sigma_{ii}}-\tfrac{1}{2}\lambda_{ir}\sigma_{ii}^{-3/2}(s_{ii}-\sigma_{ii}) \end{aligned}$$

and

$$\hat{\psi}_i^* = \hat{\psi}_i/s_{ii}$$
$$= \hat{\psi}_i/\sigma_{ii} - (\psi_i/\sigma_{ii}^2)(s_{ii}-\sigma_{ii}).$$

Since these estimates are independent of the metric of $\mathbf{x}$ we may without loss of generality assume that $\sigma_{ii} = 1$ for all values of i. Then σ_{ij} represents the correlation coefficient between x_i and x_j. With this simplification the above equations become

$$\hat{\lambda}_{ir}^* = \hat{\lambda}_{ir} - \tfrac{1}{2}\lambda_{ir}(s_{ii}-1)$$

and

$$\hat{\psi}_i^* = \hat{\psi}_i - \psi_i(s_{ii}-1).$$

Using the fact that

$$\text{cov}\,(s_{ii}, s_{jj}) = 2\sigma_{ij}^2/n,$$

we thus have

$$\text{cov}\,(\hat{\lambda}_{ir}^*, \hat{\lambda}_{jm}^*) = \text{cov}\,(\hat{\lambda}_{ir}, \hat{\lambda}_{jm}) - \tfrac{1}{2}\lambda_{jm}\,\text{cov}\,(\hat{\lambda}_{ir}, s_{jj})$$
$$-\tfrac{1}{2}\lambda_{ir}\,\text{cov}\,(\hat{\lambda}_{jm}, s_{ii}) + \tfrac{1}{2}\lambda_{ir}\lambda_{jm}\sigma_{ij}^2/n. \qquad (5.48)$$

The variance of $\hat{\lambda}_{ir}^*$ is obtained by putting $i = j$ and $r = m$ in this formula. Using (5.40) we have also

$$\text{cov}\,(\hat{\lambda}_{ir}^*, \hat{\psi}_j^*) = \text{cov}\,(\hat{\lambda}_{ir}, \hat{\psi}_j) - \psi_j\,\text{cov}\,(\hat{\lambda}_{ir}, s_{jj})$$
$$-\delta_{ij}\lambda_{ir}\psi_j^2/n + \lambda_{ir}\psi_j\sigma_{ij}^2/n. \qquad (5.49)$$

In general there is nothing to be gained by writing equations (5.48) and (5.49) more explicitly. The covariances that appear on the right-hand sides have all been found previously. However, in the case where $k = 1$ the results may be simplified considerably. For the standardised estimates $\hat{\lambda}_i^* = \hat{\lambda}_i/\sqrt{s_{ii}}$ we then have

$$\text{var}\,(\hat{\lambda}_i^*) = \text{var}\,(\hat{\lambda}_i) - \tfrac{1}{2}(1+\psi_i-2\psi_i^2)/n,$$
$$\text{cov}\,(\hat{\lambda}_i^*, \hat{\psi}_i^*) = \text{cov}\,(\hat{\lambda}_i, \hat{\psi}_i) - 2\lambda_i\psi_i^2/n,$$

and for $i \neq j$,

$$\text{cov}\,(\hat{\lambda}_i^*, \hat{\lambda}_j^*) = \text{cov}\,(\hat{\lambda}_i, \hat{\lambda}_j) - \tfrac{1}{2}\lambda_i\lambda_j(1-\psi_i\psi_j)/n,$$
$$\text{cov}\,(\hat{\lambda}_i^*, \hat{\psi}_j^*) = \text{cov}\,(\hat{\lambda}_i, \hat{\psi}_j \;-\lambda_i\lambda_j^2\psi_i\psi_j/n.$$

Returning to the general case, the covariance of $\hat{\psi}_i^*$ and $\hat{\psi}_j^*$ is, using (5.40), given by

$$n\,\text{cov}\,(\hat{\psi}_i^*, \hat{\psi}_j^*) = n\,\text{cov}\,(\hat{\psi}_i, \hat{\psi}_j) - n\psi_i\,\text{cov}\,(\hat{\psi}_j, s_{ii})$$
$$-n\psi_j\,\text{cov}\,(\hat{\psi}_i, s_{jj}) + 2\psi_i\psi_j\sigma_{ij}^2$$
$$= 2e_{ij} - 4\delta_{ij}\psi_i^3 + 2\psi_i\psi_j\sigma_{ij}^2.$$

Thus, putting $i = j$, the variance of $\hat{\psi}_i^*$ is given by

$$\text{var}\,(\hat{\psi}_i^*) = (2/n)[e_{ii}+\psi_i^2(1-2\psi_i)]. \tag{5.50}$$

For $i \neq j$ we have

$$\text{cov}\,(\hat{\psi}_i^*, \hat{\psi}_j^*) = (2/n)\,(e_{ij}+\psi_i\psi_j\sigma_{ij}^2). \tag{5.51}$$

In evaluating all expressions we put $\lambda_{ir} = \hat{\lambda}_{ir}^*$ and $\psi_i = \hat{\psi}_i^*$ for all values of i and r. We also put $\sigma_{ii} = 1$ and, for $i \neq j$,

$$\sigma_{ij} = \sum_m (\hat{\lambda}_{im}^* \hat{\lambda}_{jm}^*).$$

We may use (5.50) to estimate the standard error of $\hat{\psi}_8^*$ for data 1, which we treated in section 5.1 as if it were $\hat{\psi}_8$. We allow now for the effect of standardisation. Putting $n = 210$, $e_{88} = 2{\cdot}99$ and $\psi_8 = 0{\cdot}462$, the variance of $\hat{\psi}_8^*$ is estimated as 0·0286. Hence $\hat{\psi}_8^*$, with standard error attached, is 0·46±0·17. In this case the correction to the standard error is negligible. However, if ψ_i is near unity the correction for standardisation makes a considerable difference. In fact as $\psi_i \to 1$, $n\,\text{var}\,(\hat{\psi}_i) \to 2$, but $n\,\text{var}\,(\hat{\psi}_i^*) \to 0$.

5.4 A NUMERICAL EVALUATION OF STANDARD ERRORS OF LOADINGS

As a numerical example we give the standard errors of the loadings of *Table 4.2*, found from the analysis of data 1, first ignoring the effect of standardisation and secondly taking it into account.

The standard errors when the effect of standardisation is ignored are found as the square roots of the diagonal elements of the matrix of (5.29). They are as given in *Table 5.2*.

Table 5.2 STANDARD ERRORS OF LOADINGS OF *Table 4.2* IGNORING STANDARDISATION

	Factors		
Variate	I	II	III
1	0·066	0·058	0·076
2	0·064	0·061	0·068
3	0·070	0·071	0·083
4	0·060	0·046	0·045
5	0·065	0·057	0·072
6	0·064	0·046	0·037
7	0·068	0·057	0·066
8	0·073	0·093	0·142
9	0·066	0·047	0·055

Using the results of section 5.3 the standard errors of the loadings were recalculated, taking into account the effect of standardisation. The resulting standard errors are as shown in *Table 5.3*.

Table 5.3 STANDARD ERRORS OF LOADINGS CORRECTED FOR STANDARDISATION

Variate	Factors I	II	III
1	0·044	0·057	0·076
2	0·041	0·060	0·068
3	0·058	0·069	0·082
4	0·027	0·048	0·045
5	0·040	0·057	0·073
6	0·032	0·050	0·038
7	0·046	0·056	0·066
8	0·062	0·090	0·140
9	0·037	0·050	0·055

On comparing *Tables 5.2* and *5.3* it is clear that correcting for the effect of standardisation has in most cases considerably reduced the standard errors of the loadings on the first factor. By contrast, the standard errors of the loadings on the other two factors are little affected. This is probably due to the fact that the loadings on the first factor are, on the whole, much larger than those on the second and third factors. Standardisation imposes a ceiling of unity on the absolute value of any loading and this ceiling may be expected to reduce the variances of larger loadings more than those of smaller loadings.

Exercises

5.1. By use of (5.28) and by putting

$$\lambda_{ir} = (\theta_r - 1)^{1/2}\omega_{ir}\psi_i^{1/2},$$

show that

$$2\sum_{r=1}^{k} \lambda_{ir} b_{j\bullet ir} = \psi_i^2 g_{ij} - \delta_{ij},$$

where, as before, $g_{ij} = \phi_{ij}^2$ and

$$\phi_{ij} = (\psi_i\psi_j)^{-1/2}[\delta_{ij} - \sum_{s=1}^{k}(\omega_{is}\omega_{js})].$$

5.2. By use of (5.25) and (5.30) and by noting that

$$s_{jj} = \hat{\sigma}_{jj} = \sum_{m=1}^{k} (\hat{\lambda}_{jm}^2) + \hat{\psi}_j,$$

show that

$$n \operatorname{cov}(\hat{\lambda}_{ir}, s_{jj}) = 2 \sum_{m=1}^{k} [\lambda_{jm}(\mathbf{A} + 2\mathbf{B}'\mathbf{E}\mathbf{B})_{jm,\, ir}] + 2(\mathbf{E}\mathbf{B})_{j,\, ir}.$$

Use the result of Exercise 5.1 to verify that this is equivalent to (5.45).

5.3. Use (5.30) and the result of Exercise 5.1 to show that the covariance of $\sum_r (\hat{\lambda}_{ir}^2)$ and $\hat{\psi}_j$ is

$$2(\delta_{ij}\psi_i^2 - e_{ij})/n.$$

Hence verify that

$$\operatorname{cov}(s_{ii}, \hat{\psi}_j) = \operatorname{cov}(\hat{\sigma}_{ii}, \hat{\psi}_j) = 2\delta_{ij}\psi_i^2/n,$$

in accordance with (5.40).

5.4. With the notation

$$c_{ij} = \sum_r (\lambda_{ir}\lambda_{jr}),$$

$$d_{ij} = \sum_r [\lambda_{ir}\lambda_{jr}/(\theta_r - 1)],$$

show, using (5.47), that n times the covariance of $\sum_r (\hat{\lambda}_{ir}^2)$ and s_{jj} is equal to

$$4\sigma_{ij}(c_{ij} + d_{ij}) - 4\delta_{ij}\psi_i d_{ij} - 2(c_{ij} + d_{ij})^2 + 2d_{ij}^2,$$

which reduces to

$$2c_{ij}^2 + 4\delta_{ij}\psi_i c_{ij}$$

on putting

$$\sigma_{ij} = c_{ij} + \delta_{ij}\psi_i.$$

Hence, using (5.40), find the covariance of $\hat{\sigma}_{ii}$ and s_{jj} and verify that

$$\operatorname{cov}(\hat{\sigma}_{ii}, s_{jj}) = \operatorname{cov}(s_{ii}, s_{jj}) = 2\sigma_{ij}^2/n.$$

Chapter 6

FACTOR TRANSFORMATION AND INTERPRETATION

6.1 INTRODUCTION

Once a set of factor loadings or component weights has been found the next step is to try to interpret them in a way that gives a meaningful summary of the original data. In *Table 6.1* the correlation coefficients between the scores for a sample of 220 boys on six school subjects are given.

Table 6.1 CORRELATION COEFFICIENTS BETWEEN SIX SCHOOL SUBJECTS

1	*2*	*3*	*4*	*5*	*6*
1·000	0·439	0·410	0·288	0·329	0·248
	1·000	0·351	0·354	0·320	0·329
		1·000	0·164	0·190	0·181
			1·000	0·595	0·470
				1·000	0·464
					1·000

Two common factors were found to be adequate to account for the correlations between the variates, and the estimates of the factor loadings and of the communalities are given in *Table 6.2*.

The fact that all the correlation coefficients between the variates are positive indicates that students who get scores above average on any one of the subjects tend also to get scores above average on the other subjects. This reflection of the overall response of students to instruction, and of their relative facilities for acquiring knowledge, is variously referred to as their 'educability' or 'intelligence'. The first factor, on which all the loadings are positivel accounts for this feature of the data and might be labelled 'genera, intelligence'. The second factor represents a contrast between the first three and the last three variates, as is shown by the fact that the signs of the loadings on it are different for the two sets of variates.

A factor on which roughly half the loadings are positive and half negative is termed a bi-polar factor. (It is immaterial which pole is taken to be positive, for the signs of the loadings on the factor can be reversed without alteration of the analysis.) In our example the bi-polar factor reflects a contrast between the three verbal-type subjects, Gaelic, English and History, and the three mathematical subjects, Arithmetic, Algebra and Geometry. The variate which

Table 6.2 FACTOR LOADINGS AND COMMUNALITIES FOR SIX SCHOOL SUBJECTS

Subject	*Factor loadings* I	II	*Communality*
1. Gaelic	0·553	0·429	0·490
2. English	0·568	0·288	0·406
3. History	0·392	0·450	0·356
4. Arithmetic	0·740	−0·273	0·623
5. Algebra	0·724	−0·211	0·569
6. Geometry	0·595	−0·132	0·372

Arts v Science Good

this bi-polar factor represents is such that individuals who get above average scores on the verbal subjects tend to have above average scores on it, while individuals who get above average scores on the mathematical subjects tend to have scores below average on it. What the analysis has demonstrated is that, at a given level of general intelligence, individuals who do well on verbal subjects tend to do less well on mathematical subjects, and vice versa. By using the methods of Chapter 8 we can obtain scores on the factors, or hypothetical variates, and thus rank the individuals in order of merit on each.

Another way of looking at the data in *Table 6.2*, that is informative, is to plot the test-points in a diagram in which orthogonal axes are taken to represent the factors and the co-ordinates of the points are the factor loadings. This is done in *Figure 6.1*, and it is seen that the points appear to lie in two clusters which subtend an acute angle at the origin. The possibility then exists of interpreting the results in terms of two oblique, or correlated, factors: one 'verbal' in content and the other 'mathematical'. The general factor would then be submerged in these. Correlated factors are fairly widely used in factorial work in psychology and education and

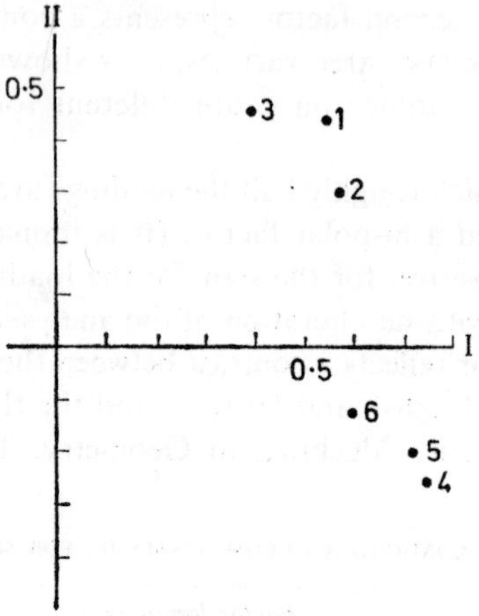

Figure 6.1

methods of estimating the correlations between them and the loadings on them will be discussed later. Before doing so it is necessary to consider the question of factor rotation.

6.2 A GRAPHICAL METHOD OF ROTATING FACTORS

Previously it has been shown that the set of loadings obtained from a factor analysis, when the number of common factors exceeds one, is not unique and that other equivalent sets can be derived by orthogonal transformations of it. This fact is frequently utilised in an effort to simplify the results of an analysis, and to make them more meaningful, even if it does allow different interpretations of the same data to be given. To illustrate the rotation procedure a further example will be valuable.

Table 6.3 gives the maximum likelihood estimates of the loadings for three factors obtained from an analysis of the correlation coefficients between the scores of a sample of 292 children on a set of 10 cognitive tests. The titles of the tests appear in the table and are more or less self-explanatory.

Since all loadings on it are positive, the first factor may, without further discussion, be taken to represent general intelligence. But when we come to factors II_0 and III_0 interpretation is not immediately obvious and is rendered difficult by the presence of a relatively large number of negative and rather small loadings. It is therefore as well to examine the results more closely and to see if they can be clarified.

The ideal way of doing this would be to make a three-dimensional model in which three orthogonal axes are taken to represent the factors. The tests could then be represented in the space by points whose co-ordinates were the loadings of the tests on the factors. Once this model has been made the actual axes employed could temporarily be forgotten and attention concentrated on the real point of interest, namely the position of the test-points relative to each other in the space. Though the axes may be rotated about their origin, or may be allowed to become oblique, the distribution

Table 6.3 FACTOR LOADINGS FOR A SET OF COGNITIVE TESTS

	Factor loadings			
Test	I_0	II_0	III_0	*Communality*
1. Comprehension	0·788	−0·152	−0·352	0·768
2. Arithmetic	0·874	0·381	0·041	0·911
3. Similarities	0·814	−0·043	−0·213	0·710
4. Vocabulary	0·798	−0·170	−0·204	0·707
5. Digit span	0·641	0·070	−0·042	0·418
6. Picture completion	0·755	−0·298	0·067	0·663
7. Picture arrangement	0·782	−0·221	0·028	0·661
8. Book design	0·767	−0·091	0·358	0·725
9. Object assembly	0·733	−0·384	0·229	0·737
10. Coding	0·771	−0·101	0·071	0·610

of the points will remain invariant. When it is examined the points may be found to lie in clusters, or perhaps to be concentrated in one or two octants of the space. If this were so it would then be reasonable to choose new axes in a way that would allow the positions of the points to be described as simply as possible, that is, using as few parameters as possible.

Since a three-dimensional model is not possible here, and in any case would be inadequate if the number of factors were greater than three, we must content ourselves by looking at the factors two at a time. When three factors are concerned three plots can be obtained, but, as a beginning, let us look at the plot of factors I_0 and III_0 in *Figure 6.2(a)*.

Examination of the figure shows that if the axes are given an orthogonal rotation in a clockwise direction, so that the first factor is made to pass through test-point 1, then all the test-points lie in the first quadrant and all the signs of the loadings become positive. In their new positions the factors are denoted by I_1 and III_1 and it is clear that I_1 can still be labelled 'general intelligence'; for test 1, as its name implies, is a test of comprehension. To obtain the

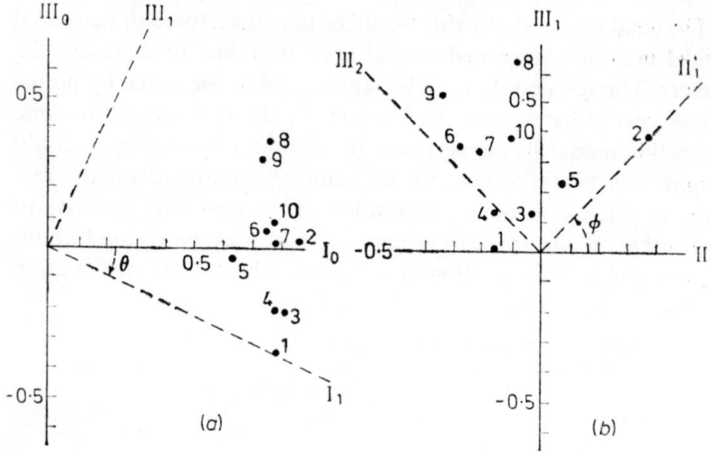

Figure 6.2

loadings on these new axes or factors the original 10×2 matrix of loadings for factors I_0 and III_0 is post-multiplied by an orthogonal matrix of the form

$$\begin{bmatrix} \cos\theta & \sin\theta \\ -\sin\theta & \cos\theta \end{bmatrix}.$$

The angle θ can be determined either by measurement from *Figure 6.2(a)* or by calculation. It is about 24° and the new loadings on factors I_1 and III_1 are given in *Table 6.4(a)*.

Further plots can now be made using the loadings on I_1, II_0 and III_1. That for factors II_0 and III_1, where subscripts refer to the number of rotations a factor has had, is given in *Figure 6.2(b)*. Again it is clear that an orthogonal rotation of the axes, this time in an anticlockwise direction, can be performed which will eliminate nearly all negative loadings. (All can be eliminated by a further slight rotation of I_1 and II_1.) Here it is convenient to let the axis of factor II pass through test point 2, that is the Arithmetic test, as this may help in the interpretation of this new factor. Let us denote the rotation matrix now required by

$$\begin{bmatrix} \cos\phi & -\sin\phi \\ \sin\phi & \cos\phi \end{bmatrix}.$$

The angle ϕ in the figure is about 46° and the new loadings on the factors, denoted by II_1 and III_2, together with the loadings on I_1, are given in *Table 6.4(b)*. In eliminating almost all negative signs from the loadings we have shown that the three orthogonal axes representing the factors can be so placed as to include all the test points in one octant of the factor space. However, we must hasten to add that other positions of the axes, different (though perhaps only slightly so) from those given here, would also have achieved this end.

Table 6.4 LOADINGS ON ROTATED FACTORS

	(a) First rotation		*(b) Second rotation*		
Test	I_1	III_1	I_1	II_1	III_2
1	0·863	0·000	0·863	−0·106	0·109
2	0·781	0·394	0·781	0·548	0·000
3	0·830	0·138	0·830	0·069	0·127
4	0·812	0·139	0·812	0·018	0·219
5	0·602	0·223	0·602	0·209	0·105
6	0·662	0·369	0·662	0·058	0·471
7	0·703	0·345	0·703	0·094	0·399
8	0·554	0·640	0·554	0·397	0·510
9	0·576	0·508	0·576	0·098	0·629
10	0·675	0·379	0·675	0·202	0·363

Unhampered by the presence of many negative signs we can now look at the new set of loadings in *Table 6.4(b)* and see if the factors can be interpreted in a meaningful way. Factor I_1 can, with reasonable conviction, still be described as a factor of 'general intelligence'. On factor III_2 the largest loadings are for the last five tests of the set. The resemblance between these tests, after the effect of factor I_1 has been removed, is thought to be concerned with the ability to visualise; so this third factor could tentatively be labelled 'visualisation'. On factor II_1 the largest loadings are for Arithmetic and Block design. Since the latter test is not concerned with numbers it would be unwise to think of the factor merely as one of numerical ability, and a better description of it might be 'mechanical knowledge', or even 'memory'.

The example just given illustrates a common type of approach to the problem of interpreting the results of factorial studies of an exploratory nature. Clearly the subjective element in it is large and, as a consequence, different investigators might interpret the same data differently. This possibility, however natural, is generally thought to be undesirable and to avoid it various empirical techniques for transforming factors, which in some predetermined sense lead to a unique set of results, have been proposed. Of these the most commonly used are the *varimax method* for orthogonal rotation due to Kaiser (1958), and the *promax method* for transformation to correlated factors due to Hendrickson and White (1964). In addition, methods will be described for transforming factors in such a way that the final loadings are in reasonable conformity with a given *pattern* in which some of the elements are postulated to be zero (Lawley and Maxwell, 1964). Here we may emphasise that all these transformation procedures are in essence approximate and exploratory in nature. Their primary use should be to enable the experimenter to formulate hypotheses about the possible factor content of his variables which can be submitted to verification on further data by a confirmatory type of analysis (for which see Chapter 7).

6.3 THE VARIMAX METHOD

In using the graphic method of rotation the main objects are to eliminate negative loadings and to describe the data by as few non-zero loadings as possible. The aims of the varimax method are somewhat similar. The factors are rotated in such a way that the new loadings tend to be either relatively large or relatively small in absolute magnitude compared with the original ones. This is accomplished, as we shall see, by maximising a certain function of the squares of the loadings, the procedure being iterative. In Kaiser's original method the factors were rotated in pairs until the loadings converged to their final values. More recent variations of the method are discussed in a book by Horst (1965, Chapter 18). One of Horst's procedures, which he calls 'the simultaneous factor varimax solution' is described below. The main feature of this is that in each iteration all factors are rotated simultaneously. It has the advantages of being faster than the original method and of not accumulating rounding errors.

We shall denote by $\mathbf{\Lambda}_0$ the $p \times k$ matrix of unrotated loadings. The ith column of $\mathbf{\Lambda}_0'$, a vector of order k, is denoted by $\mathbf{l}_i$. Let $\mathbf{M}$ be an orthogonal rotation matrix of order k whose rth column is $\mathbf{m}_r$ $(r = 1, \ldots, k)$. The matrix $\mathbf{\Lambda} = [\lambda_{ir}]$ of rotated loadings is given by

$$\mathbf{\Lambda} = \mathbf{\Lambda}_0\mathbf{M},$$

and thus

$$\lambda_{ir} = \mathbf{l}_i'\mathbf{m}_r.$$

We define the scalars d_r $(r = 1, \ldots, k)$ by

$$d_r = \sum_{i=1}^{p} \lambda_{ir}^2 = \sum_i (\mathbf{l}_i'\mathbf{m}_r)^2. \tag{6.1}$$

Thus d_r is the sum of squares of the loadings in the rth column of $\mathbf{\Lambda}$.

The criterion x that is maximised in the simultaneous varimax method is given by

$$x = \sum_{r=1}^{k}\left[\sum_{i=1}^{p}(\lambda_{ir}^2 - d_r/p)^2\right]$$
$$= \sum_{r}\left[\sum_{i}\lambda_{ir}^4 - d_r^2/p\right]$$
$$= \sum_{r}\left[\sum_{i}(\mathbf{l}_i'\mathbf{m}_r)^4 - d_r^2/p\right]. \qquad (6.2)$$

From this definition it is clear that x represents the sum of squares of deviations of the values of λ_{ir}^2, each measured as a deviation from the corresponding column mean d_r/p. The maximisation is with respect to the elements of $\mathbf{M}$. Since the columns of $\mathbf{M}$ satisfy the conditions

$$\mathbf{m}_r'\mathbf{m}_r = 1,$$
$$\mathbf{m}_r'\mathbf{m}_s = 0 \quad (r \neq s),$$

we must equate to zero the derivatives with respect to the elements of $\mathbf{M}$ of the expression

$$y = x - 2\sum_{r}\sum_{s} a_{rs}\mathbf{m}_r'\mathbf{m}_s, \qquad (6.3)$$

where the a-coefficients are indeterminate multipliers such that a_{sr} is identical with a_{rs}.

By use of (6.1), (6.2) and (6.3) we find that the derivative of y with respect to the vector $\mathbf{m}_s$ is given by

$$\partial y/\partial \mathbf{m}_s = 4\sum_{i}[(\mathbf{l}_i'\mathbf{m}_s)^3\mathbf{l}_i] - (4d_s/p)\sum_{i}[(\mathbf{l}_i'\mathbf{m}_s)\mathbf{l}_i] - 4\sum_{r}(a_{rs}\mathbf{m}_r)$$
$$= 4\sum_{i}(c_{is} - d_s\lambda_{is}/p)\mathbf{l}_i - 4\sum_{r}a_{rs}\mathbf{m}_r,$$

where

$$c_{ir} = (\mathbf{l}_i'\mathbf{m}_r)^3 = \lambda_{ir}^3.$$

This expression for $\partial y/\partial \mathbf{m}_s$ is found on examination to be the sth column of the matrix

$$4[\mathbf{\Lambda}_0'\mathbf{C} - (1/p)\mathbf{\Lambda}_0'\mathbf{\Lambda}\mathbf{D} - \mathbf{M}\mathbf{A}],$$

where $\mathbf{C}$ is the $p \times k$ matrix with elements c_{ir}, $\mathbf{A}$ is the symmetric matrix of order k with elements a_{rs} and $\mathbf{D}$ is the diagonal matrix whose diagonal elements are $d_1, \ldots, d_k$. Hence, taking all values of s into account, we have

$$\partial y/\partial \mathbf{M} = 4(\mathbf{B} - \mathbf{M}\mathbf{A}),$$

where

$$\mathbf{B} = \mathbf{\Lambda}_0'[\mathbf{C} - (1/p)\mathbf{\Lambda}\mathbf{D}]. \qquad (6.4)$$

The orthogonal matrix $\mathbf{M}$ that maximises x thus satisfies the equation

$$\mathbf{MA} = \mathbf{B}. \tag{6.5}$$

It can be shown that if $\mathbf{A}$ is positive definite as well as being symmetric, the above equation corresponds to a *maximum* for x. On pre-multiplying (6.5) by $\mathbf{M}'$ we have

$$\begin{aligned}\mathbf{A} &= \mathbf{M'B} \\ &= \mathbf{\Lambda'C} - (1/p)\mathbf{\Lambda'\Lambda D}.\end{aligned}$$

Thus the matrix on the right-hand side is symmetric when x is maximised. The rth diagonal element of $\mathbf{A}$ is

$$\sum_i \lambda_{ir}^4 - (1/p)\,(\sum_i \lambda_{ir}^2)^2.$$

From the definition of x in (6.2) it follows that tr ($\mathbf{A}$) is its maximum value.

The matrices $\mathbf{M}$, $\mathbf{A}$ and $\mathbf{B}$ that satisfy (6.5) are found by an iterative procedure. We start with an initial approximation $\mathbf{M}_1$. This yields an initial approximation $\mathbf{\Lambda}_1 = \mathbf{\Lambda}_0\mathbf{M}_1$ for $\mathbf{\Lambda}$, and from $\mathbf{\Lambda}_1$ we obtain approximations $\mathbf{C}_1$, $\mathbf{D}_1$ and $\mathbf{B}_1$ for $\mathbf{C}$, $\mathbf{D}$ and $\mathbf{B}$ respectively. We now have to find a symmetric and positive definite matrix $\mathbf{A}_1$ and an orthogonal matrix $\mathbf{M}_2$ satisfying the equation

$$\mathbf{M}_2\mathbf{A}_1 = \mathbf{B}_1. \tag{6.6}$$

Note that on pre-multiplying each side of this equation by its transpose we have

$$\mathbf{A}_1^2 = \mathbf{B}_1'\mathbf{B}_1.$$

Now $\mathbf{B}_1'\mathbf{B}_1$ is a symmetric and positive definite matrix of order k. Hence we may express it in the form $\mathbf{U\Delta^2U'}$, where $\mathbf{U}$ is orthogonal and $\mathbf{\Delta}$ is diagonal with positive diagonal elements. We are then able to find $\mathbf{A}_1$ in the form

$$\mathbf{A}_1 = \mathbf{U\Delta U'},$$

and from (6.6) $\mathbf{M}_2$ is given by

$$\mathbf{M}_2 = \mathbf{B}_1\mathbf{A}_1^{-1}.$$

The procedure is repeated with $\mathbf{M}_2$ in place of $\mathbf{M}_1$ and, if necessary, with a sequence of matrices $\mathbf{M}_s$. It can be shown that for $\mathbf{M}_1$ sufficiently near to $\mathbf{M}$ this sequence converges to $\mathbf{M}$. The corresponding sequence of matrices $\mathbf{A}_s$ is such that the values of tr ($\mathbf{A}_s$) form an ascending sequence of values converging to the maximum of the criterion x. In practice a good initial approximation to $\mathbf{M}$ is unnecessary, so that it is usual to take $\mathbf{M}_1 = \mathbf{I}$ and $\mathbf{\Lambda}_1 = \mathbf{\Lambda}_0$. Very few iterations are as a rule required.

Apart from the notation the above description differs in one small respect from that given by Horst. He suggests that, before the maximisation process begins, the matrix $\mathbf{\Lambda}_0$ should be standardised by rows. Thus each row vector $\mathbf{l}_i'$ is scaled to have unit length. At the end of the maximisation, when **M** has been found, the final $\mathbf{\Lambda}$ must be rescaled by rows so that the communalities are given their correct values. We have adopted this modification in the second of the two numerical examples that follow.

The varimax method appears to have two main limitations. In the first place the pattern of loadings obtained by its use may change considerably if additional factors are included in the rotation. In the second place the method is not very helpful in cases where a dominant general factor exists. For example when the method was applied to the loadings of *Table 6.3* (without standardisation by rows) it gave the results shown in *Table 6.5*. These compare rather poorly with the results of *Table 6.4(b)* obtained by graphic rotation. It is worth noting, however, that the highest loadings on each factor correspond to the same variates in both rotations. But in cases where a dominant general factor exists the experimenter always has the option of omitting it from the rotation.

Table 6.5 VARIMAX ROTATION OF THE LOADINGS OF *Table 6.3*

Variate	*Factor loadings* I	II	III	*Communality*
1	0·759	0·329	0·289	0·768
2	0·340	0·849	0·274	0·911
3	0·633	0·450	0·327	0·710
4	0·657	0·345	0·397	0·707
5	0·370	0·453	0·276	0·418
6	0·464	0·263	0·615	0·663
7	0·485	0·332	0·561	0·660
8	0·183	0·472	0·684	0·725
9	0·354	0·209	0·754	0·737
10	0·409	0·423	0·514	0·610

Rotation matrix **M**

0·560	0·633	0·534
−0·311	0·758	−0·573
−0·768	0·155	0·622

In the absence of a dominant general factor a varimax rotation is generally useful as a first step in interpreting the data. When the method was applied to the matrix of loadings given in *Table 4.6*, in this case after standardising by rows, it gave the results shown in *Table 6.6*. In this table the names of the variates have been entered. The first five are tests of cognitive ability and the last five are scales measuring orectic tendency.

Table 6.6 VARIMAX ROTATION OF THE LOADINGS IN *Table 4.6*

	Factor loadings			
Variates	I	II	III	IV
Verbal ability	0·027	0·591	0·307	0·413
Spacial ability	0·040	0·095	0·122	0·593
Reasoning	0·059	0·416	0·198	0·695
Number	0·110	0·330	0·079	0·485
Verbal fluency	0·060	0·765	0·072	0·244
Neuroticism	0·229	0·109	0·357	0·174
Ways to be different	0·151	0·240	0·780	0·160
Worries and anxieties	0·926	0·043	0·370	0·058
Interests	0·272	0·108	0·438	0·180
Annoyances	0·089	0·002	0·625	0·045

After rotation it is clear that factors II and IV are cognitive factors; factor II is one of high verbal content, while factor IV is more concerned with reasoning, numerical and spatial ability. On the other hand, factors I and III are orectic factors; factor I separates 'worries and anxieties' from the other four scales.

6.4 THE PROMAX METHOD

In many cases the pattern of loadings may be still further simplified by transforming to oblique, or correlated, factors. Such a simplification is, however, achieved at a price since the correlation coefficients between these factors must be estimated, and they must be taken into consideration in any interpretation of the results. The problem of making a transformation to oblique factors was first considered by Thurstone (1947, Chapters 15 et seq.). The graphic methods used by him were also lucidly described by Thomson (1951). Unfortunately they involved the experimenter in many arbitrary decisions about the choice of factors. As a consequence attention was eventually directed to the discovery of analytic criteria that could be used to produce unique and less subjective solutions. The earliest work on this problem was done by Carroll (1953) and by Kaiser and Dickman (1959), but here we shall confine our attention to a method given by Hendrickson and White

(1964), called the promax method, which seems to work well in practice.

The promax method starts with a varimax rotation of the original loadings. We suppose that this produces a $p \times k$ loading matrix $\mathbf{\Lambda} = [\lambda_{ir}]$. From it is constructed another $p \times k$ matrix $\mathbf{Q} = [q_{ir}]$ whose elements are defined by

$$q_{ir} = |\lambda_{ir}^{m-1}| \lambda_{ir}, \tag{6.7}$$

where m is some integer satisfying $m > 1$. Thus $q_{ir} = 0$ if $\lambda_{ir} = 0$; otherwise q_{ir} has the same sign as λ_{ir} and has the same absolute value as λ_{ir}^m. We now seek a transformation matrix $\mathbf{U}$, not necessarily orthogonal, such that for $r = 1, \ldots, k$ the rth column of $\mathbf{\Lambda U}$ is in maximum conformity in a least-squares sense with the rth column of $\mathbf{Q}$. When $\mathbf{U}$ has been found the effect of its use is, in general, to increase in absolute magnitude loadings that are relatively large and decrease those that are small. The value chosen for m is a matter of trial and error, but experience has shown that values greater than 4, while they may simplify the pattern of loadings, often produce factors that are very highly correlated; this is undesirable in practice.

Let us denote the rth columns of $\mathbf{Q}$ and $\mathbf{U}$ by $\mathbf{q}_r$ and $\mathbf{u}_r$ respectively. Then for each value of r we choose $\mathbf{u}_r$ such that it minimises the expression

$$(\mathbf{q}_r - \mathbf{\Lambda u}_r)' (\mathbf{q}_r - \mathbf{\Lambda u}_r).$$

The derivative of this with respect to the vector $\mathbf{u}_r$ is

$$-2\mathbf{\Lambda}'(\mathbf{q}_r - \mathbf{\Lambda u}_r).$$

This is equated to zero to give

$$(\mathbf{\Lambda}'\mathbf{\Lambda})\mathbf{u}_r = \mathbf{\Lambda}'\mathbf{q}_r.$$

Hence, taking all values of r into account, we have

$$(\mathbf{\Lambda}'\mathbf{\Lambda})\mathbf{U} = \mathbf{\Lambda}'\mathbf{Q},$$

or

$$\mathbf{U} = (\mathbf{\Lambda}'\mathbf{\Lambda})^{-1}\mathbf{\Lambda}'\mathbf{Q}. \tag{6.8}$$

In practice it is convenient to rescale the columns of $\mathbf{U}$ so that the transformed factors have unit variances. This is accomplished by finding the diagonal matrix $\mathbf{D}$, with positive diagonal elements, that satisfies

$$\mathbf{D}^2 = \text{diag}\,[(\mathbf{U}'\mathbf{U})^{-1}], \tag{6.9}$$

where diag $(\mathbf{X})$ denotes the diagonal part of $\mathbf{X}$. The transformation matrix $\mathbf{M}$, that replaces $\mathbf{U}$, is then given by

$$\mathbf{M} = \mathbf{UD}, \tag{6.10}$$

and the matrix of transformed loadings is

$$\mathbf{\Lambda}^* = \mathbf{\Lambda M}. \tag{6.11}$$

We now have

$$\begin{aligned}\mathbf{\Lambda\Lambda}' &= \mathbf{\Lambda}^*\mathbf{M}^{-1}\mathbf{M}'^{-1}\mathbf{\Lambda}^{*\prime} \\ &= \mathbf{\Lambda}^*(\mathbf{M}'\mathbf{M})^{-1}\mathbf{\Lambda}^{*\prime} \\ &= \mathbf{\Lambda}^*\mathbf{\Phi}\mathbf{\Lambda}^{*\prime},\end{aligned}$$

where

$$\mathbf{\Phi} = (\mathbf{M}'\mathbf{M})^{-1} = \mathbf{D}^{-1}(\mathbf{U}'\mathbf{U})^{-1}\mathbf{D}^{-1}. \tag{6.12}$$

A reference to equation (2.3) of Chapter 2 makes it clear that $\mathbf{\Phi}$ is the covariance matrix for the transformed factors. From the definition of $\mathbf{D}$ in (6.9) it is evident that $\mathbf{\Phi}$ has unit diagonal elements. Hence $\mathbf{\Phi}$ is also the correlation matrix for the new factors, which have in fact been standardised.

The promax method, with $m = 2$, was applied to the varimax loadings of *Table 6.5*. The results are shown in *Table 6.7*.

Table 6.7 PROMAX TRANSFORMATION OF THE LOADINGS IN *Table 6.5*

Variate	*Factor loadings* I	II	III
1	0·737	0·139	0·086
2	−0·012	0·924	0·074
3	0·522	0·320	0·137
4	0·590	0·163	0·230
5	0·207	0·410	0·141
6	0·362	0·060	0·539
7	0·364	0·155	0·460
8	−0·086	0·388	0·646
9	0·234	0·011	0·732
10	0·236	0·304	0·408

Transformation matrix **M**			*Factor correlations*		
1·222	−0·279	−0·239	1·000	0·621	0·461
−0·459	1·277	−0·183		1·000	0·481
−0·136	−0·238	1·134			1·000

The pattern of loadings in *Table 6.7* is much simpler than that in *Table 6.5*, and there is a substantial number of very small loadings. This would be a help in setting up a hypothesis of the type considered in Chapter 7.

6.5 TRANSFORMATION OF CORRELATED FACTORS USING A PATTERN MATRIX

In using the methods described in the last two sections the hope is that the factors obtained will be readily interpretable. However, the interpretation of factors depends ultimately on prior knowledge of the nature of the variates being analysed, and the experimenter may naturally wish to use this knowledge when rotating or transforming the factors. In this section and the next we describe methods that are appropriate if it is possible to postulate a factor pattern in which the positions of zero or near-zero loadings are specified in advance. These methods were given by Lawley and Maxwell (1964).

We discuss first the transformation of a set of orthogonal factors to new factors that are allowed to be correlated. We shall denote the original $p \times k$ matrix of loadings by $\mathbf{\Lambda} = [\lambda_{ir}]$ and the new matrix by $\mathbf{\Lambda}^* = [\lambda^*_{ir}]$. The experimenter is required to supply a $p \times k$ pattern matrix $\mathbf{Q} = [q_{ir}]$, each of whose elements is either 0 or 1. If λ^*_{ir} is a 'restricted' loading, whose value has to be zero or small, then $q_{ir} = 0$; otherwise $q_{ir} = 1$.

The new loadings are given by $\mathbf{\Lambda}^* = \mathbf{\Lambda U}$, where $\mathbf{U}$ is a non-singular $k \times k$ transformation matrix. The problem is to choose $\mathbf{U}$ in such a way that $\mathbf{\Lambda}^*$ does in fact have zero or small elements in positions corresponding to zero elements of $\mathbf{Q}$. We shall assume that $\mathbf{Q}$ is such that $\mathbf{U}$ and $\mathbf{\Lambda}^*$ can be uniquely determined. It is not possible to give general conditions for uniqueness that are both necessary and sufficient. A necessary condition is that each column of $\mathbf{Q}$ should contain at least $k-1$ zero elements; but the positions as well as the numbers of these elements are important.

Let $\mathbf{u}_r$ denote the rth column of $\mathbf{U}$. Then $\boldsymbol{\lambda}^*_r$, the rth column of $\mathbf{\Lambda}^*$, is given by

$$\boldsymbol{\lambda}^*_r = \mathbf{\Lambda u}_r,$$

and the sum of squares of its elements is

$$\boldsymbol{\lambda}^{*\prime}_r \boldsymbol{\lambda}^*_r = \mathbf{u}'_r(\mathbf{\Lambda}'\mathbf{\Lambda})\mathbf{u}_r. \tag{6.13}$$

Suppose that for any given value of r, from 1 to k, we replace *all* elements in the ith row of $\mathbf{\Lambda}$ (or in the ith column of $\mathbf{\Lambda}'$) by zeros, for $i = 1, \ldots, p$, whenever $q_{ir} = 0$. This produces a matrix that we denote by $\mathbf{\Lambda}_r$. As a simple example suppose that, with $p = 7$ and $k = 3$, the rth row of $\mathbf{Q}'$ is

$$[1 \quad 0 \quad 1 \quad 0 \quad 0 \quad 1 \quad 1].$$

Then, with crosses denoting non-zero elements, $\mathbf{\Lambda}_r'$ takes the form:

$$\begin{matrix} \times & 0 & \times & 0 & 0 & \times & \times \\ \times & 0 & \times & 0 & 0 & \times & \times \\ \times & 0 & \times & 0 & 0 & \times & \times \end{matrix}$$

Now consider the vector $\mathbf{\Lambda}_r\mathbf{u}_r$. This is seen on examination to be the result of replacing all restricted elements of $\boldsymbol{\lambda}_r^*$ by zeros. Hence the sum of squares of all non-restricted elements of $\boldsymbol{\lambda}_r^*$ is

$$(\mathbf{\Lambda}_r\mathbf{u}_r)'(\mathbf{\Lambda}_r\mathbf{u}_r) = \mathbf{u}_r'(\mathbf{\Lambda}_r'\mathbf{\Lambda}_r)\mathbf{u}_r. \tag{6.14}$$

Let us choose $\mathbf{u}_r$ in such a way that the sum of squares of the restricted elements of $\boldsymbol{\lambda}_r^*$ is minimised, subject to the sum of squares of all elements being held constant. This is equivalent to maximising the sum of squares of the non-restricted elements. Thus we must maximise the expression (6.14) subject to that of (6.13) remaining constant. To accomplish this we equate to zero the derivative with respect to the vector $\mathbf{u}_r$ of the expression

$$\mathbf{u}_r'(\mathbf{\Lambda}_r'\mathbf{\Lambda}_r)\mathbf{u}_r - \alpha_r\mathbf{u}_r'(\mathbf{\Lambda}'\mathbf{\Lambda})\mathbf{u}_r,$$

where α_r is an indeterminate multiplier. This yields the equation

$$(\mathbf{\Lambda}_r'\mathbf{\Lambda}_r)\mathbf{u}_r = \alpha_r(\mathbf{\Lambda}'\mathbf{\Lambda})\mathbf{u}_r, \tag{6.15}$$

or

$$\mathbf{H}_r\mathbf{u}_r = \alpha_r\mathbf{u}_r, \tag{6.16}$$

where

$$\mathbf{H}_r = (\mathbf{\Lambda}'\mathbf{\Lambda})^{-1}(\mathbf{\Lambda}_r'\mathbf{\Lambda}_r). \tag{6.17}$$

An inspection of equation (6.16) makes it clear that α_r is a latent root of the $k\times k$ matrix $\mathbf{H}_r$ and that $\mathbf{u}_r$ is the corresponding latent column vector. On pre-multiplying (6.15) by $\mathbf{u}_r'$ we find that α_r is the ratio of $\mathbf{u}_r'(\mathbf{\Lambda}_r'\mathbf{\Lambda}_r)\mathbf{u}_r$ to $\mathbf{u}_r'(\mathbf{\Lambda}'\mathbf{\Lambda})\mathbf{u}_r$. Since this ratio is to be maximised, we must take α_r to be the largest latent root of $\mathbf{H}_r$. In general this will be a distinct root. It can be shown that if the rth column of $\mathbf{Q}$ has exactly $k-1$ zeros, then $\alpha_r = 1$. If there are more than $k-1$ zeros, then $0 < \alpha_r \leq 1$; but in practice α_r is usually very near to unity.

Since rapid and powerful methods exist for finding the latent roots and vectors of *symmetric* matrices, there is an advantage in calculating the vector $\mathbf{u}_r$ by the following method. Let $\mathbf{T}$ be a lower triangular matrix with positive diagonal elements such that

$\mathbf{TT}' = \mathbf{\Lambda}'\mathbf{\Lambda}$ and let $\mathbf{W}_r$ be the matrix defined by

$$\mathbf{W}_r = \mathbf{T}^{-1}(\mathbf{\Lambda}_r'\mathbf{\Lambda}_r)\mathbf{T}'^{-1}. \tag{6.18}$$

Then $\mathbf{W}_r$ is a symmetric and positive definite matrix having the same latent roots as $\mathbf{H}_r$. Let $\mathbf{v}_r$ be the latent vector of $\mathbf{W}_r$ corresponding to the largest latent root α_r. Then we have

$$\mathbf{u}_r = \mathbf{T}'^{-1}\mathbf{v}_r. \tag{6.19}$$

The vectors $\mathbf{u}_r$ are calculated successively for $r = 1, \ldots, k$. Their scales and signs are arbitrary. In order that each of the transformed factors should have unit variance we rescale the $\mathbf{u}_r$ in the same way as in section 6.4. We find the diagonal matrix $\mathbf{D}$, with positive diagonal elements, given by (6.9). Then in place of $\mathbf{U}$ we use $\mathbf{M} = \mathbf{UD}$ as a transformation matrix. The sign of each column of $\mathbf{M}$ may be altered if necessary to ensure that each new factor has predominantly positive loadings. The new loading matrix is given by $\mathbf{\Lambda}^* = \mathbf{\Lambda M}$ and the factor correlation matrix is $\mathbf{\Phi} = (\mathbf{M}'\mathbf{M})^{-1}$.

If the original loading matrix $\mathbf{\Lambda}$ has been estimated by the maximum likelihood method of Chapter 4, there is some advantage in substituting $\mathbf{\Psi}^{-1/2}\mathbf{\Lambda}$ for $\mathbf{\Lambda}$ throughout, and similarly for $\mathbf{\Lambda}_r$, where $\mathbf{\Psi}$ is as usual the diagonal matrix with the residual variances as its diagonal elements. Since the matrix

$$\mathbf{\Delta} = \mathbf{\Lambda}'\mathbf{\Psi}^{-1}\mathbf{\Lambda} = \mathbf{TT}'$$

is diagonal, we have $\mathbf{T} = \mathbf{T}' = \mathbf{\Delta}^{1/2}$, and equation (6.18) now becomes

$$\mathbf{W}_r = \mathbf{\Delta}^{-1/2}(\mathbf{\Lambda}_r'\mathbf{\Psi}^{-1}\mathbf{\Lambda}_r)\mathbf{\Delta}^{-1/2}. \tag{6.20}$$

In place of (6.19) we have

$$\mathbf{u}_r = \mathbf{\Delta}^{-1/2}\mathbf{v}_r. \tag{6.21}$$

As a numerical illustration we apply the above method to the results given in *Table 4.2*. Since these are maximum likelihood estimates we use the modification of the preceding paragraph. The transpose $\mathbf{Q}'$ of the pattern matrix is given by

$$\mathbf{Q}' = \begin{bmatrix} 1 & 1 & 1 & 1 & 1 & 0 & 1 & 0 & 1 \\ 0 & 0 & 0 & 1 & 1 & 1 & 0 & 0 & 0 \\ 1 & 0 & 0 & 1 & 0 & 1 & 1 & 1 & 1 \end{bmatrix}.$$

The new loadings, the transformation matrix and the factor correlations are as shown in *Table 6.8*. The restricted loadings are given in brackets; they are all either very small or zero.

Table 6.8 NON-ORTHOGONAL TRANSFORMATION OF THE LOADINGS OF *Table 4.2*

Variate	Factor loadings I	II	III
1	0·495	(0·044)	0·337
2	0·691	(0·101)	(0·035)
3	0·668	(−0·059)	(−0·050)
4	0·201	0·724	0·178
5	0·228	0·663	(0·002)
6	(0·000)	0·811	0·316
7	0·689	(−0·047)	0·177
8	(0·000)	(0·003)	0·734
9	0·703	(−0·015)	0·288

Transformation matrix **M**			*Factor correlations*		
0·472	0·527	0·311	1·000	0·429	0·506
0·774	−0·978	0·148		1·000	0·135
−0·907	0·106	1·116			1·000

6.6 ROTATION TO NEW ORTHOGONAL FACTORS USING A PATTERN MATRIX

The experimenter may wish to rotate to new orthogonal factors using a pattern matrix. We give a method that will usually be suitable in this situation. A necessary, though not sufficient, condition for the rotation to be unique is that the pattern matrix **Q** should now have at least $\frac{1}{2}k(k-1)$ zeros altogether. In addition we assume that the columns of **Q** can be and have been ordered in such a way that, for $r = 1, \ldots, k-1$, there are at least $k-r$ zeros in the rth column. (The last column need have no zeros.) This assumption is necessary for the rotation method to work. In some cases there may be more than one way of so ordering the columns of **Q**, but the choice of order will not as a rule affect the final outcome very much.

The factor rotation is carried out in $k-1$ stages. In the first stage $\mathbf{\Lambda}_1$, $\mathbf{W}_1$, $\mathbf{v}_1$ and $\mathbf{u}_1$ are found exactly as in the previous section. But $\mathbf{m}_1$ is now simply the result of sealing $\mathbf{u}_1$ to unit length; it satisfies $\mathbf{m}_1'\mathbf{m}_1 = 1$. Next we construct an orthogonal rotation matrix $\mathbf{M}_1$, of order k, having $\mathbf{m}_1$ as its first column (see section A1.17 of Appendix I). All k factors are then rotated to produce a new loading matrix $\mathbf{\Lambda M}_1$. The first column of this, namely $\mathbf{\Lambda m}_1$, is the first column of $\mathbf{\Lambda}^*$.

If $k > 2$, another rotation is required. We delete the first column of $\mathbf{\Lambda M}_1$ and rename the resulting $p\times(k-1)$ matrix $\mathbf{\Lambda}$. From this we find $\mathbf{\Lambda}_2$, $\mathbf{W}_2$, $\mathbf{v}_2$ and $\mathbf{u}_2$ in the same way as before. The matrix $\mathbf{W}_2$ is of order $k-1$ and the vectors $\mathbf{v}_2$ and $\mathbf{u}_2$ each contain $k-1$ elements. By sealing $\mathbf{u}_2$ to unit length we obtain $\mathbf{m}_2$. After constructing an orthogonal matrix $\mathbf{M}_2$ whose first column is $\mathbf{m}_2$ we rotate all $k-1$ remaining factors to obtain a new loading matrix $\mathbf{\Lambda M}_2$. The first column of this, namely $\mathbf{\Lambda m}_2$, forms the second column of $\mathbf{\Lambda}^*$.

If $k > 3$, the procedure continues. At the rth stage $\mathbf{\Lambda}$ and $\mathbf{\Lambda}_r$ are $p\times(k-r+1)$ matrices and $\mathbf{W}_r$ is of order $k-r+1$. The last $k-r+1$ factors are rotated by the use of the orthogonal matrix $\mathbf{M}_r$, and $\mathbf{\Lambda m}_r$ forms the rth column of $\mathbf{\Lambda}^*$. At the final stage, when $r = k-1$, the last two factors are rotated and $\mathbf{\Lambda M}_{k-1}$ provides the last two columns of $\mathbf{\Lambda}^*$.

As a numerical illustration of this method we apply it to the results of *Table 4.2*. This time the pattern matrix is given by

$$\mathbf{Q}' = \begin{bmatrix} 0 & 0 & 0 & 1 & 1 & 1 & 0 & 0 & 0 \\ 1 & 0 & 0 & 1 & 0 & 1 & 1 & 1 & 1 \\ 1 & 1 & 1 & 1 & 1 & 1 & 1 & 1 & 1 \end{bmatrix}.$$

In finding $\mathbf{W}_1$ we made use of equation (6.20). In the second stage of the rotation, however, we used (6.18) to obtain $\mathbf{W}_2$ with $\mathbf{\Lambda}_2'\mathbf{\Psi}^{-1}\mathbf{\Lambda}_2$ in place of $\mathbf{\Lambda}_2'\mathbf{\Lambda}_2$ and with $\mathbf{T}$ satisfying

$$\mathbf{TT}' = \mathbf{\Lambda}'\mathbf{\Psi}^{-1}\mathbf{\Lambda},$$

since at this stage the matrix on the right-hand side is not diagonal.

The new loadings and the two rotation matrices are as shown in *Table 6.9*.

Table 6.9 ORTHOGONAL ROTATION OF THE LOADINGS OF *Table 4.2*

Variate	*Factor loadings* I	II	III
1	(0·038)	0·311	0·672
2	(0·090)	(0·049)	0·751
3	(−0·054)	(−0·014)	0·617
4	0·647	0·108	0·598
5	0·594	(−0·044)	0·513
6	0·726	0·215	0·501
7	(−0·043)	0·186	0·751
8	(0·003)	0·645	0·349
9	(−0·015)	0·281	0·831

	Rotation matrices			
	M_1		M_2	
0·471	0·860	0·198	0·491	0·871
−0·877	0·481	0·000	−0·871	0·491
0·095	0·173	−0·980		

Exercises

6.1. Apply the varimax method of section 6.3, with standardisation by rows, to the loadings of *Table 4.2*. Show that the resulting rotated loadings are as follows:

I	II	III
0·573	0·264	0·389
0·661	0·342	0·137
0·594	0·163	0·062
0·319	0·813	0·160
0·280	0·736	0·004
0·189	0·851	0·251
0·691	0·217	0·277
0·243	0·144	0·683
0·743	0·269	0·380

6.2. Apply the promax method of section 6.4, with $m = 3$, to the loadings given in Exercise 6.1. Compare the results with those of *Table 6.8*.

6.3. By use of the transformation method of section 6.5 with pattern matrix

$$\begin{bmatrix} 1 & 1 & 1 & 0 & 0 & 0 \\ 0 & 0 & 0 & 1 & 1 & 1 \end{bmatrix}$$

find the loadings of the 6 variates of *Table 6.2* on two correlated factors and estimate the correlation coefficient between them. Since the loadings in this table were estimated by the maximum likelihood method of Chapter 4, the modifications of (6.20) and (6.21) may be used. (The residual variance of each variate is estimated by subtracting its communality from unity.) Compare the results obtained with those given in analysis *(a)* of Exercise 7.4 of Chapter 7.

6.4. By use of the rotation method of section 6.6 with pattern matrix

$$\begin{bmatrix} 1 & 1 & 1 & 1 & 1 & 1 \\ 1 & 1 & 1 & 0 & 0 & 0 \end{bmatrix}$$

find the loadings of the 6 variates of *Table 6.2* on two orthogonally rotated factors. The modifications of (6.20) and (6.21) may be used to find $\mathbf{W}_1$ and $\mathbf{u}_1$. Compare the results obtained with those of analysis *(b)* of Exercise 7.4. In this case there is very close agreement.

Chapter 7

ESTIMATION IN RESTRICTED FACTOR MODELS

7.1 INTRODUCTION

In this chapter we consider the problem of estimating parameters in the model of section 2.5 of Chapter 2, in which the factors, more than one in number, may be correlated. We suppose that, possibly by previous use of the methods of Chapter 6, the experimenter is able to set up a hypothesis that defines the parameters uniquely. The hypothesis H must specify not only the number of factors but also the values of certain elements in $\mathbf{\Lambda}$ and in $\mathbf{\Phi}$, the factor covariance matrix. It may in addition specify the values of some or all of the residual variances ψ_i, though this would be unusual in practice. As a rule the values specified for the elements of $\mathbf{\Lambda}$ or for the non-diagonal elements of $\mathbf{\Phi}$ are zeros, but other values could be used. We shall always suppose that the factors are standardised and hence that the diagonal elements of $\mathbf{\Phi}$ are, by hypothesis, unities. All models of the above kind will in future be described as 'restricted' models.

Let the numbers of specified or fixed elements in $\mathbf{\Lambda}$, $\mathbf{\Phi}$ and $\mathbf{\Psi}$ be denoted respectively by n_λ, n_ϕ and n_ψ. The total number m of fixed parameters is then given by

$$m = n_\lambda + n_\phi + n_\psi.$$

The total number of parameters in $\mathbf{\Lambda}$, $\mathbf{\Phi}$ and $\mathbf{\Psi}$ is

$$pk + \tfrac{1}{2}k(k+1) + p = \tfrac{1}{2}(2p+k)(k+1).$$

Hence the total number of unspecified or free parameters is

$$\tfrac{1}{2}(2p+k)(k+1) - m.$$

For the hypothesis H to be non-trivial this number must be less than $\frac{1}{2}p(p+1)$, the number of distinct elements of $\mathbf{\Sigma}$. This requirement is equivalent to the inequality

$$p^2+m > \tfrac{1}{2}(p+k)(p+k+1). \qquad (7.1)$$

A necessary, though not sufficient, condition for the parameters to be uniquely defined is that

$$n_\lambda + n_\phi \geqslant k^2.$$

The free parameters are to be estimated by the method of maximum likelihood. Methods whose aim was to maximise the likelihood for restricted models were first given by Howe (1955), Anderson and Rubin (1956) and Lawley (1958). In all these methods partial derivatives with respect to the free parameters were found and equated to zero. After some algebraical simplification iterative procedures for solving the equations were suggested. Recent work has shown, however, that these procedures do not always converge. Even when convergence does occur it is usually very slow. A better method, for which ultimate convergence is assured, was given by Jöreskog (1966b); but it was still difficult in some cases to obtain a very accurate solution unless many iterations were performed. Efficient maximisation of the likelihood function seems impossible without the use of second-order derivatives.

The latest method for performing restricted maximum likelihood factor analysis, which we give in subsequent sections, has been described in papers by Lawley (1967) and by Jöreskog (1969). See also a more expository paper by Jöreskog and Lawley (1968). A computer program for the method has been written by Jöreskog and Gruvaeus (1967) and appears to be highly successful. The generality of the method gives it great flexibility since it will deal with a wide variety of hypotheses. For example, no essential distinction is made between models in which the factors are correlated and those with orthogonal factors. It is possible to have hybrid cases in which one group of factors may be correlated while the remaining factors are assumed to be uncorrelated with this group or with one another.

As with the unrestricted model of Chapter 4, it is more convenient in practice to minimise a certain function F of the unknown parameters than to maximise the likelihood. The Fletcher and Powell method is again used to carry out the minimisation. Unfortunately a two-stage procedure such as that formerly employed is no longer possible except in special cases. The function F has therefore to be minimised simultaneously with respect to all free parameters. The $\mathbf{E}$ matrix evaluated in each iteration converges finally to the inverse of the matrix of second-order derivatives with respect to the free parameters. Thus in some cases its order may be quite large. If, for example, there were 40 variates and 10 common factors, the number of free parameters, which is the order of **E**, might well be almost 400. The inversion and storage of matrices whose order is as large as this presents considerable difficulties. However, with the development of computers having greater storage capacity than those of today, these difficulties may in time disappear.

7.2 THE FUNCTION *F* AND ITS DERIVATIVES

As in section 4.2 we shall assume that the vector variate **x** has a multivariate normal distribution and, as before, we shall use **S** to denote the sample covariance matrix obtained from a random sample of $N(>p)$ sets of observations. Then the likelihood function L based only on the information provided by **S** is, omitting a function of the observations, given by

$$\log_e L = -\tfrac{1}{2}n\,[\log_e|\mathbf{\Sigma}|+\mathrm{tr}\,(\mathbf{S\Sigma}^{-1})],$$

exactly as in equation (4.2), with $n = N-1$. Now, however, $\mathbf{\Sigma}$ is given by

$$\mathbf{\Sigma} = \mathbf{\Lambda\Phi\Lambda}'+\mathbf{\Psi}, \tag{7.2}$$

as in equation (2.3).

Instead of maximising L it is more convenient to minimise the function $F = F(\mathbf{\Lambda},\mathbf{\Phi},\mathbf{\Psi})$ defined by

$$F(\mathbf{\Lambda},\mathbf{\Phi},\mathbf{\Psi}) = \log|\mathbf{\Sigma}|+\mathrm{tr}\,(\mathbf{S\Sigma}^{-1})-\log|\mathbf{S}|-p. \tag{7.3}$$

The minimisation is with respect to the free parameters. All other parameters are kept fixed at the values specified by the hypothesis H.

The derivatives of F with respect to the free parameters are found by the same methods as those used in section 4.2. In differentiating with respect to $\mathbf{\Phi} = [\phi_{rs}]$ we have to remember that this is a symmetric matrix of order k with $\frac{1}{2}k(k+1)$ distinct elements. We have

$$\partial F/\partial\mathbf{\Lambda} = 2\mathbf{\Sigma}^{-1}(\mathbf{\Sigma}-\mathbf{S})\mathbf{\Sigma}^{-1}\mathbf{\Lambda\Phi}, \tag{7.4}$$

$$\partial F/\partial\mathbf{\Phi} = 2\mathbf{\Lambda}'\mathbf{\Sigma}^{-1}(\mathbf{\Sigma}-\mathbf{S})\mathbf{\Sigma}^{-1}\mathbf{\Lambda}, \tag{7.5}$$

$$\partial F/\partial\mathbf{\Psi} = \mathrm{diag}\,[\mathbf{\Sigma}^{-1}(\mathbf{\Sigma}-\mathbf{S})\mathbf{\Sigma}^{-1}], \tag{7.6}$$

with the understanding that the elements of the three matrices on the left that correspond in position to fixed elements of $\mathbf{\Lambda},\mathbf{\Phi}$ and $\mathbf{\Psi}$ respectively are taken to be zero. In particular the diagonal elements of $\partial F/\partial\mathbf{\Phi}$ are zero. The computation of the above matrices is simplified by making use of the identities

$$\mathbf{\Sigma}^{-1} = \mathbf{\Psi}^{-1} - \mathbf{\Psi}^{-1}\mathbf{\Lambda}(\mathbf{I}+\mathbf{\Phi}\mathbf{\Gamma})^{-1}\mathbf{\Phi}\mathbf{\Lambda}'\mathbf{\Psi}^{-1} \tag{7.7}$$

and

$$\mathbf{\Sigma}^{-1}\mathbf{\Lambda} = \mathbf{\Psi}^{-1}\mathbf{\Lambda}(\mathbf{I}+\mathbf{\Phi}\mathbf{\Gamma})^{-1}, \tag{7.8}$$

where

$$\mathbf{\Gamma} = \mathbf{\Lambda}'\mathbf{\Psi}^{-1}\mathbf{\Lambda}.$$

The matrix $\mathbf{\Gamma}$ is not, in general, diagonal.

For computational purposes it is convenient to arrange the free parameters in $\mathbf{\Lambda}, \mathbf{\Phi}$ and $\mathbf{\Psi}$ in the form of a column vector $\boldsymbol{\theta}$ (or row vector $\boldsymbol{\theta}'$). This may be done as follows. Let $\boldsymbol{\theta}_r$, for $r = 1, \ldots, k$, be a vector containing the free parameters in the rth column of $\mathbf{\Lambda}$, let $\boldsymbol{\theta}_{k+1}$ be a vector containing the free parameters in $\mathbf{\Phi}$ in some suitable order and let $\boldsymbol{\theta}_{k+2}$ be a vector containing the free parameters in $\mathbf{\Psi}$. (If the factors are orthogonal, $\boldsymbol{\theta}_{k+1}$ has no elements.) Then $\boldsymbol{\theta}'$ is the composite vector $(\boldsymbol{\theta}_1' \; \boldsymbol{\theta}_2' \ldots \boldsymbol{\theta}_{k+2}')$. We may now regard F as a function of the elements of $\boldsymbol{\theta}$ and write it as $F(\boldsymbol{\theta})$. We carry out a similar arrangement of all first-order derivatives of F that have not been set equal to zero. These form a vector that we denote as $\partial F/\partial\boldsymbol{\theta}$. Let q be the total number of free parameters. Then $\boldsymbol{\theta}$ and $\partial F/\partial\boldsymbol{\theta}$ each have q elements.

In the minimisation procedure it is necessary to compute repeatedly not only the derivatives but also the value of the function F, as given by (7.3). In this expression the matrix $\mathbf{\Sigma}^{-1}$ is found by use of (7.7). We require also to evaluate $|\mathbf{\Sigma}|$. By making use of result (A1.9) of Appendix I we have

$$\begin{aligned} |\mathbf{\Sigma}| &= |\mathbf{\Psi}+\mathbf{\Lambda}\mathbf{\Phi}\mathbf{\Lambda}'| \\ &= |\mathbf{\Psi}|\,|\mathbf{I}_p+\mathbf{\Psi}^{-1}\mathbf{\Lambda}\mathbf{\Phi}\mathbf{\Lambda}'| \\ &= |\mathbf{\Psi}|\,|\mathbf{I}_k+\mathbf{\Phi}\mathbf{\Lambda}'\mathbf{\Psi}^{-1}\mathbf{\Lambda}| \\ &= (\psi_1\psi_2\ldots\psi_p)\,|\mathbf{I}+\mathbf{\Phi}\mathbf{\Gamma}|. \end{aligned} \tag{7.9}$$

The exact expressions for the second-order derivatives of F with respect to the parameters are somewhat complicated in form [see equation (A2.9) of Appendix II]. However, it is easy to find initial approximations to them that are sufficiently good for our purposes. We shall henceforth assume that n is reasonably large.

As an example, consider the evaluation of $\partial^2 F/\partial\lambda_{ir}\partial\lambda_{js}$, where λ_{ir} and λ_{js} are both free parameters. From (7.4) we have

$$\partial F/\partial\lambda_{ir} = 2[\mathbf{\Sigma}^{-1}(\mathbf{\Sigma}-\mathbf{S})\mathbf{\Sigma}^{-1}\mathbf{\Lambda}\mathbf{\Phi}]_{ir}. \tag{7.10}$$

Differentiating this with respect to λ_{js} we have, approximately,

$$\partial^2 F/\partial\lambda_{ir}\partial\lambda_{js} = 2[\mathbf{\Sigma}^{-1}(\partial\mathbf{\Sigma}/\partial\lambda_{js})\mathbf{\Sigma}^{-1}\mathbf{\Lambda}\mathbf{\Phi}]_{ir}. \tag{7.11}$$

The other terms that would arise on the right-hand side may be neglected since the elements of the matrix $(\mathbf{\Sigma}-\mathbf{S})$ are small in probability. Since also $E(\mathbf{\Sigma}-\mathbf{S}) = \mathbf{0}$, the above is an exact expression for $E(\partial^2 F/\partial\lambda_{ir}\partial\lambda_{js})$. Now

$$\begin{aligned}\partial\mathbf{\Sigma}/\partial\lambda_{js} &= (\partial/\partial\lambda_{js})(\mathbf{\Lambda\Phi\Lambda}'+\mathbf{\Psi})\\ &= (\partial\mathbf{\Lambda}/\partial\lambda_{js})\mathbf{\Phi\Lambda}'+\mathbf{\Lambda\Phi}(\partial\mathbf{\Lambda}/\partial\lambda_{js})',\end{aligned}$$

and $\partial\mathbf{\Lambda}/\partial\lambda_{js}$ is a $p\times k$ matrix whose elements are all zero except for that in the jth row and sth column, which is unity. Hence, substituting for $\partial\mathbf{\Sigma}/\partial\lambda_{js}$ in (7.11), we have

$$E(\partial^2 F/\partial\lambda_{ir}\partial\lambda_{js}) = 2\sigma^{ij}(\mathbf{\Phi\Lambda}'\mathbf{\Sigma}^{-1}\mathbf{\Lambda\Phi})_{sr}+2(\mathbf{\Sigma}^{-1}\mathbf{\Lambda\Phi})_{is}(\mathbf{\Sigma}^{-1}\mathbf{\Lambda\Phi})_{jr}, \quad (7.12)$$

where σ^{ij} is, as before, the element in the ith row and jth column of $\mathbf{\Sigma}^{-1}$.

An alternative method of deriving the above expression is to use the result, well-known in estimation theory, that

$$E(\partial^2 \log_e L/\partial\lambda_{ir}\lambda_{js}) = -E[(\partial \log_e L/\partial\lambda_{ir})(\partial \log_e L/\partial\lambda_{js})].$$

Since $\log_e L$ is $-\frac{1}{2}n$ times F plus a function of the observations only, this is equivalent to

$$E(\partial^2 F/\partial\lambda_{ir}\partial\lambda_{js}) = \tfrac{1}{2}nE[(\partial F/\partial\lambda_{ir})(\partial F/\partial\lambda_{js})].$$

The right-hand side of this equation may be evaluated by the methods of section 5.1 of Chapter 5, with $\partial F/\partial\lambda_{ir}$ as given in (7.10) and with a similar expression for $\partial F/\partial\lambda_{js}$. This yields the result given in (7.12).

The expectations of other second-order derivatives are found in a similar manner. They may all be expressed concisely in terms of the elements of various matrices. Let ξ_{ir} and η_{ir} denote the elements in the ith row and rth column of the respective $p\times k$ matrices $\mathbf{\Sigma}^{-1}\mathbf{\Lambda}$ and $\mathbf{\Sigma}^{-1}\mathbf{\Lambda\Phi}$. Define the symmetric matrix $\mathbf{A}$ of order k by

$$\mathbf{A} = \mathbf{\Lambda}'\mathbf{\Sigma}^{-1}\mathbf{\Lambda} = \mathbf{\Gamma}(\mathbf{I}+\mathbf{\Phi\Gamma})^{-1}.$$

Let α_{rs}, β_{rs} and μ_{rs} be the elements in the rth row and sth column of the respective matrices $\mathbf{A}, \mathbf{\Phi A}$ and $\mathbf{\Phi A\Phi}$. Then the required results are as follows:

$$\begin{aligned}E(\partial^2 F/\partial\lambda_{ir}\partial\lambda_{js}) &= 2(\sigma^{ij}\mu_{rs}+\eta_{is}\eta_{jr}),\\ E(\partial^2 F/\partial\lambda_{ir}\partial\phi_{st}) &= 2(\xi_{is}\beta_{rt}+\xi_{it}\beta_{rs}),\\ E(\partial^2 F/\partial\lambda_{ir}\partial\psi_{j}) &= 2\sigma^{ij}\eta_{jr},\\ E(\partial^2 F/\partial\phi_{rs}\partial\phi_{tu}) &= 2(\alpha_{rt}\alpha_{su}+\alpha_{ru}\alpha_{st}),\\ E(\partial^2 F/\partial\phi_{rs}\partial\psi_{i}) &= 2\xi_{ir}\xi_{is},\\ E(\partial^2 F/\partial\psi_{i}\partial\psi_{j}) &= (\sigma^{ij})^2.\end{aligned}$$

In the above expressions it is understood that differentiation is with respect to the free parameters only. The expectations of the second-order derivatives, suitably arranged, thus form a symmetric matrix **G** of order q. Since n is assumed to be moderately large, **G** may be used to approximate the derivatives themselves, provided that the value of each parameter does not differ too much from its maximum likelihood estimate. Subject to this proviso, **G** when evaluated should prove to be positive definite. The arrangement of the rows and columns of **G** must correspond to the arrangement of the free parameters within the vector $\boldsymbol{\theta}$. Corresponding to the partitioning of $\boldsymbol{\theta}$ into the sub-vectors $\boldsymbol{\theta}_1, \ldots, \boldsymbol{\theta}_{k+2}$ we have a partitioning of the matrix **G**. Thus

$$\mathbf{G} = \begin{bmatrix} \mathbf{G}_{11} & \mathbf{G}_{12} & \cdot & \cdot & \cdot & \mathbf{G}_{1,k+2} \\ \mathbf{G}_{21} & \mathbf{G}_{22} & \cdot & \cdot & \cdot & \mathbf{G}_{2,k+2} \\ \cdot & \cdot & \cdot & \cdot & \cdot & \cdot \\ \cdot & \cdot & & & & \cdot \\ \cdot & \cdot & & & & \cdot \\ \mathbf{G}_{k+2,1} & \mathbf{G}_{k+2,2} & \cdot & \cdot & \cdot & \mathbf{G}_{k+2,k+2} \end{bmatrix}.$$

The sub-matrix $\mathbf{G}_{uv} = \mathbf{G}'_{vu}$ may be represented symbolically as $\partial^2 F/\partial\boldsymbol{\theta}_u \partial\boldsymbol{\theta}'_v$.

7.3 THE MINIMISATION PROCEDURE

The procedure for minimising F begins with initial estimates of all the free parameters in $\boldsymbol{\theta}$. The better these are the fewer iterations will be required. If the fixed parameters consist mainly of zero loadings and perhaps also of zero factor correlations, the methods of Chapter 6 may be used to provide the initial estimates. Thus, for example, we might first carry out the unrestricted maximum likelihood procedure of Chapter 4 and then transform or rotate the loadings by means of the methods of sections 6.5 and 6.6. In the RMLFA program of Jöreskog and Gruvaeus initial estimates are generated by a modified centroid method, which we need not here describe, if these have not been provided by the user.

From the initial estimate of $\boldsymbol{\theta}$ we are able to compute the function value and the vector $\partial F/\partial\boldsymbol{\theta}$ of first derivatives. Experience has shown that it is usually best to begin with a few steepest descent iterations rather than to use second-order derivatives from the start, since if **G** is evaluated using the initial estimates, it may be very inaccurate and possibly not even positive definite. For the remaining iterations, when one is comparatively near to the minimum

7*

of F, the Fletcher and Powell method is employed. For this purpose an initial $\mathbf{E}$ matrix is required. If q, the number of free parameters, is not too large, this is obtained by inverting the matrix $\mathbf{G}$ of the preceding section.

When q is large the inversion of $\mathbf{G}$ may be difficult and time-consuming. In such cases it is better to replace all sub-matrices $\mathbf{G}_{uv}$ for which $u \neq v$ by null matrices. This is reasonable since previous studies of sampling variability have shown that, for $u \neq v$, the estimates of parameters in $\boldsymbol{\theta}_u$ are often only slightly correlated with those in $\boldsymbol{\theta}_v$. Hence $\mathbf{G}$ is replaced by the matrix $\mathbf{G}^*$ given by

$$\mathbf{G}^* = \begin{bmatrix} \mathbf{G}_{11} & \mathbf{0} & . & . & . & \mathbf{0} \\ \mathbf{0} & \mathbf{G}_{22} & . & . & . & \mathbf{0} \\ . & . & & & & . \\ . & . & & & & . \\ . & . & & & & . \\ \mathbf{0} & \mathbf{0} & . & . & . & \mathbf{G}_{k+2,\,k+2} \end{bmatrix}.$$

This is easily inverted to provide the initial $\mathbf{E}$ matrix.

The iterative procedure converges fairly rapidly. The final vector $\hat{\boldsymbol{\theta}}$ has as its elements the maximum likelihood estimates of the free parameters. By standard estimation theory the asymptotic variances and covariances of these are estimated by the matrix $(2/n)\mathbf{E}$. To calculate this asymptotic covariance matrix we could use the final $\mathbf{E}$ matrix given by the Fletcher and Powell method. It is, however, theoretically, better, to recalculate $\mathbf{G}$, using $\hat{\boldsymbol{\theta}}$, and then to find $\mathbf{E}$ by inversion.

7.4 TESTS OF HYPOTHESES

The likelihood ratio criterion λ for testing the hypothesis H may be found in the same way as in section 4.4 of Chapter 4. As in (4.26) we have

$$-2 \log_e \lambda = n[\log_e |\hat{\boldsymbol{\Sigma}}| + \operatorname{tr}(\mathbf{S}\hat{\boldsymbol{\Sigma}}^{-1}) - \log_e |\mathbf{S}| - p], \quad (7.13)$$

but now

$$\hat{\boldsymbol{\Sigma}} = \hat{\boldsymbol{\Lambda}}\hat{\boldsymbol{\Phi}}\hat{\boldsymbol{\Lambda}}' + \hat{\boldsymbol{\Psi}}.$$

In the matrices $\hat{\boldsymbol{\Lambda}}$, $\hat{\boldsymbol{\Phi}}$ and $\hat{\boldsymbol{\Psi}}$ maximum likelihood estimates are substituted for all free parameters, while other parameters take their specified values.

When n is large, $-2 \log \lambda$ is distributed under H approximately as χ^2. The number of degrees of freedom is equal to $\frac{1}{2}p(p+1)$, the number of distinct elements of $\boldsymbol{\Sigma}$, minus q, the number of free parameters. Using m, as in section 7.1, to denote the number of fixed parameters, we find that the number of degrees of freedom for χ^2 is given by

$$\begin{aligned} d &= \tfrac{1}{2}p(p+1)-\tfrac{1}{2}(2p+k)(k+1)+m \\ &= p^2-\tfrac{1}{2}(p+k)(p+k+1)+m. \end{aligned} \tag{7.14}$$

This number is positive provided that inequality (7.1) is satisfied.

The expression of (7.13) cannot in general be simplified further since, except in special cases, tr $(\mathbf{S}\hat{\boldsymbol{\Sigma}}^{-1})$ is not equal to p. However, the expression is simply the minimum value of F and is computed as $F(\hat{\boldsymbol{\theta}})$. The χ^2 approximation may be further improved if in (7.13) n is replaced by a suitable multiplying factor. The exact value for this is at present unknown. In the circumstances it seems best to use $n-(2p+5)/6$, which is the appropriate factor for the trivial case where $k = 0$. Hence as an approximate χ^2 criterion for testing H we shall use the statistic

$$U = [n-(2p+5)/6]F(\hat{\boldsymbol{\theta}}). \tag{7.15}$$

The number of degrees of freedom is as given in (7.14).

7.5 NUMERICAL EXAMPLES

To illustrate the methods of this chapter we again make use of data 1 of Chapter 4. However, variate 2 has been omitted and the remaining variates have been rearranged in the order 5, 4, 6, 8, 9, 7, 1, 3. The resulting standardised covariance matrix is as shown in *Table 7.1*.

Table 7.1 COVARIANCE MATRIX FOR 8 VARIATES

1	*2*	*3*	*4*	*5*	*6*	*7*	*8*
1·000	0·691	0·679	0·149	0·409	0·382	0·346	0·270
	1·000	0·791	0·285	0·505	0·443	0·471	0·355
		1·000	0·314	0·472	0·372	0·426	0·254
			1·000	0·470	0·385	0·434	0·218
				1·000	0·680	0·639	0·504
					1·000	0·576	0·452
						1·000	0·395
							1·000

We first set up the hypothesis that, with 3 correlated factors, the pattern matrix for the loadings is as follows:

$$\begin{matrix} 1 & 1 & 0 & 0 & 1 & 1 & 1 & 1 \\ 1 & 1 & 1 & 0 & 0 & 0 & 0 & 0 \\ 0 & 1 & 1 & 1 & 1 & 1 & 1 & 0 \end{matrix}$$

Here a zero indicates that the corresponding element of $\mathbf{\Lambda}'$ is specified to be zero, while a unity indicates that the element is unspecified. Thus the total number of fixed parameters, including the unit diagonal elements of $\mathbf{\Phi}$, is 12.

Extremely crude initial estimates of the free parameters were used. In spite of this the minimum value of F was found correct to five significant figures after 28 iterations, including 5 steepest descent iterations at the beginning. None of the final first-order derivatives exceeded 0·0001 in absolute value. The maximum likelihood estimates of the parameters are as given in *Table 7.2*. The ith column of $\hat{\mathbf{\Lambda}}$ is denoted by $\hat{\boldsymbol{\lambda}}_i$ and $\hat{\mathbf{\Psi}}$ is given in column form.

Table 7.2 MAXIMUM LIKELIHOOD SOLUTION WITH THREE CORRELATED FACTORS

Variate	$\hat{\boldsymbol{\lambda}}_1$	$\hat{\boldsymbol{\lambda}}_2$	$\hat{\boldsymbol{\lambda}}_3$	$\hat{\mathbf{\Psi}}$
1	0·353	0·644	0	0·352
2	0·268	0·676	0·241	0·224
3	0	0·769	0·451	0·144
4	0	0	0·651	0·576
5	0·601	0	0·369	0·242
6	0·663	0	0·187	0·380
7	0·441	0	0·400	0·438
8	0·600	0	0	0·640

Factor correlation matrix $\hat{\mathbf{\Phi}}$

1·000	0·237	0·587
	1·000	0·087
		1·000

The minimum value of F is given by $F(\hat{\boldsymbol{\theta}}) = 0{\cdot}024772$. We have $n = 210$, $p = 8$ and $k = 3$. Hence the multiplying factor in (7.15) is 206·5, giving $U = 5{\cdot}12$. Putting $m = 12$, we find from (7.14) that the number of degrees of freedom for this approximate χ^2 criterion is 10. Since the value of U is well below expectation the hypothesis that we set up is accepted.

This example is not intended to be realistic. In fact we are not justified in using a test of significance here, since the hypothesis was formulated only after a preliminary examination of the data. Thus it is not surprising that it was accepted. Using the same covariance matrix as before, let us now set up the hypothesis that, with three orthogonal factors, the pattern matrix for $\mathbf{\Lambda}'$ is as follows:

$$\begin{matrix} 1 & 1 & 1 & 1 & 1 & 1 & 1 & 1 \\ 1 & 1 & 1 & 0 & 0 & 0 & 0 & 0 \\ 0 & 1 & 1 & 1 & 1 & 1 & 1 & 0 \end{matrix}$$

All elements of $\mathbf{\Phi}$ are fixed; those in the diagonal are specified to be unity, as before, while the non-diagonal elements are all zero. Hence the total number of fixed parameters is now 13.

Once again very crude initial estimates of the free parameters were used. This time 22 iterations were required to achieve about the same degree of accuracy as before. The maximum likelihood estimates of the parameters are given in *Table 7.3*.

Table 7.3 MAXIMUM LIKELIHOOD SOLUTION WITH THREE ORTHOGONAL FACTORS

Variate	$\hat{\boldsymbol{\lambda}}_1$	$\hat{\boldsymbol{\lambda}}_2$	$\hat{\boldsymbol{\lambda}}_3$	$\hat{\boldsymbol{\Psi}}$
1	0·499	0·626	0	0·359
2	0·562	0·658	0·179	0·221
3	0·436	0·741	0·340	0·149
4	0·366	0	0·541	0·573
5	0·809	0	0·321	0·242
6	0·767	0	0·175	0·381
7	0·665	0	0·345	0·438
8	0·605	0	0	0·634

In this example $\text{tr}\,(\mathbf{S}\hat{\mathbf{\Sigma}}^{-1}) \neq p$ and some of the diagonal elements of $\hat{\mathbf{\Sigma}}$ are not exactly equal to the corresponding elements of S, as is the case with unrestricted maximum likelihood estimation.

The minimum value of F is given by $F(\hat{\boldsymbol{\theta}}) = 0{\cdot}025098$. The multiplying factor in (7.15) is 206·5, as before; so the test criterion is 5·18. Since $m = 13$, the number of degrees of freedom for U is 11. The value of U is again well below expectation and the hypothesis is accepted.

7.6 USE OF DATA TO FORMULATE AND TO TEST A HYPOTHESIS

The examples of the preceding section were intended mainly as illustrations of the computational aspects of the estimation procedure. From a theoretical point of view it is obviously unjustifiable to use the same data both to set up a hypothesis and to test it. To redress the balance we now give an example in which half of the data is used to carry out an exploratory factor analysis and to formulate a hypothesis. A confirmatory analysis is then performed on the other half of the data and the hypothesis is tested.

Some data of Holzinger and Swineford (1939) are used. The analysis given below is taken from a paper by Jöreskog and Lawley (1968). The data consist of various psychological tests administered to seventh and eighth grade children in two different schools. Only the Grant–White sample is used here. This sample was divided into two samples of sizes 73 and 72 respectively, referred to in future as the exploration sample and the confirmation sample. A detailed description of the tests is given in Holzinger and Swineford's paper. The following nine test variates have been selected, the numbers in parentheses being their original code numbers: 1. Visual perception (1), 2. Cubes (2), 3. Lozenges (4), 4. Paragraph comprehension (6), 5. Sentence completion (7), 6. Word meaning (9), 7. Addition (10), 8. Counting dots (12), 9. Straight-curved capitals (13). The two correlation matrices were computed directly from the test scores and are as given in *Table 7.4*. In this table correlation coefficients above the diagonal refer to the exploration sample and those below it to the confirmation sample.

The correlation matrix of the exploration sample was first analysed, using the unrestricted maximum likelihood method of esti-

Table 7.4 CORRELATION COEFFICIENTS FOR EXPLORATION AND CONFIRMATION

1	*2*	*3*	*4*	*5*	*6*	*7*	*8*	*9*
1·000	0·411	0·479	0·401	0·370	0·393	0·078	0·389	0·411
0·245	1·000	0·463	0·223	0·198	0·244	−0·042	0·169	0·324
0·418	0·362	1·000	0·231	0·272	0·357	−0·126	0·153	0·307
0·282	0·217	0·425	1·000	0·659	0·688	0·215	0·221	0·256
0·257	0·125	0·304	0·784	1·000	0·649	0·293	0·279	0·324
0·239	0·131	0·330	0·743	0·730	1·000	0·226	0·298	0·294
0·122	0·149	0·265	0·185	0·221	0·118	1·000	0·602	0·446
0·253	0·183	0·329	0·021	0·139	−0·027	0·601	1·000	0·630
0·583	0·147	0·455	0·381	0·400	0·235	0·385	0·462	1·000

mation. The hypotheses that the number of factors is 0, 1, 2 and 3 were successively tested. The results of these tests are shown in *Table 7.5*. For each value of k the value of the test criterion U_k,

Table 7.5 TESTS OF HYPOTHESES FOR THE EXPLORATION SAMPLE

k	U_k	d_k	P
0	253·00	36	<0·001
1	101·30	27	<0·001
2	46·39	19	<0·001
3	5·14	12	0·95

the number of degrees of freedom d_k and the corresponding probability P are given. From the table it is clear that three factors are needed to account adequately for the correlations.

The maximum likelihood solution for three factors is as given in *Table 7.6*. To obtain a preliminary interpretation of the data the factors were rotated orthogonally by means of the varimax method. The resulting varimax loadings are as shown in *Table 7.7*. Since the sample size is rather small, sampling variability is very large. Hence only factor loadings larger than 0·30 in absolute magnitude are considered to be meaningful. The first factor, which has its largest loadings in variates 1, 2, 3 and 9 appears to be a visual factor. It seems also that the second factor, determined by variates 4, 5 and 6, is a verbal factor and that the third factor, determined by variates 7, 8 and 9, is a speed factor.

Table 7.6 UNRESTRICTED MAXIMUM LIKELIHOOD SOLUTION FOR EXPLORATION SAMPLE

Variate	$\hat{\lambda}_1$	$\hat{\lambda}_2$	$\hat{\lambda}_3$	$\hat{\Psi}$
1	0·59	−0·14	0·37	0·49
2	0·37	−0·19	0·45	0·62
3	0·42	−0·32	0·53	0·44
4	0·71	−0·37	−0·27	0·29
5	0·71	−0·26	−0·23	0·37
6	0·74	−0·33	−0·17	0·33
7	0·50	0·58	−0·30	0·32
8	0·65	0·54	0·13	0·27
9	0·64	0·34	0·27	0·40

Table 7.7 VARIMAX ROTATED LOADINGS FOR EXPLORATION SAMPLE

Variate	*Factor loadings* I	II	III
1	0·61	0·30	0·23
2	0·60	0·12	0·06
3	0·72	0·18	−0·02
4	0·17	0·82	0·11
5	0·18	0·75	0·21
6	0·26	0·76	0·16
7	−0·22	0·22	0·76
8	0·21	0·12	0·82
9	0·40	0·14	0·65

To carry further our interpretation of the data let us apply the method of section 6.5 to the results in *Table 7.6*. The factors are then transformed to a new set of correlated factors. The pattern matrix for the loadings, transposed for convenience, is as follows:

$$\begin{matrix} 1 & 1 & 1 & 0 & 0 & 0 & 0 & 0 & 1 \\ 0 & 0 & 0 & 1 & 1 & 1 & 0 & 0 & 0 \\ 0 & 0 & 0 & 0 & 0 & 0 & 1 & 1 & 1 \end{matrix}$$

The new loadings and the factor correlations are given in *Table 7.8*. The results are evidently a refinement of those of *Table 7.7*. With

Table 7.8 NON-ORTHOGONAL TRANSFORMATION OF THE LOADINGS OF *Table 7.6*

Variate	*Factor loadings* *Visual*	*Verbal*	*Speed*
1	0·60	0·14	0·14
2	0·63	−0·02	0·00
3	0·74	0·03	−0·11
4	−0·02	0·88	−0·06
5	0·01	0·77	0·05
6	0·10	0·78	0·00
7	−0·24	0·14	0·78
8	0·26	−0·09	0·82
9	0·44	−0·08	0·63

	Factor correlations *Visual*	*Verbal*	*Speed*
Visual	1·00	0·45	0·13
Verbal		1·00	0·36
Speed			1·00

a few exceptions the large loadings have become larger and the small loadings have become smaller.

The analysis of the correlation matrix of the exploration sample suggests the hypothesis that the nine test variates depend upon three correlated factors with the loading matrix **Λ** having the pattern given above. In the pattern matrix a zero now indicates that the corresponding loading is specified to be zero, while a unity indicates that the corresponding loading is to be estimated from the data. The confirmation sample is used to test the hypothesis, after applying the estimation method of this chapter. The maximum likelihood estimates under the hypothesis are given in *Table 7.9*.

Table 7.9 RESTRICTED MAXIMUM LIKELIHOOD SOLUTION FOR CONFIRMATION SAMPLE

Variate	$\hat{\boldsymbol{\lambda}}_1$	$\hat{\boldsymbol{\lambda}}_2$	$\hat{\boldsymbol{\lambda}}_3$	$\hat{\boldsymbol{\Psi}}$
1	0·68	0	0	0·54
2	0·34	0	0	0·88
3	0·66	0	0	0·57
4	0	0·91	0	0·18
5	0	0·87	0	0·25
6	0	0·82	0	0·32
7	0	0	0·65	0·58
8	0	0	0·92	0·15
9	0·67	0	0·19	0·39
		Factor correlation matrix $\hat{\boldsymbol{\Phi}}$		
	1·00	0·55	0·47	
		1·00	0·09	
			1·00	

As usual the factors are made to have unit variances; so the diagonal elements of **Φ** are specified to be unities. Hence the total number of fixed parameters is 20. The test criterion U is distributed under the hypothesis approximately as χ^2 with 23 degrees of freedom. Its value is 28·28, which corresponds to a probability of 0·20 and is thus not significant at the 5 per cent level. The hypothesis is therefore accepted.

There is a recognisable measure of agreement between the loadings of *Tables 7.8* and *7.9*, though in one or two cases corresponding values differ considerably. The factor correlations in the two tables do not resemble each other very closely. The differences are, however, mainly attributable to the large sampling errors that arise from the use of fairly small samples.

In this example various other, slightly different, hypotheses would no doubt fit the data equally well. Some of these have been discussed and tested by Jöreskog (1969), using the undivided sample of size 145. In Jöreskog's paper the model is a little more general than ours in that the diagonal elements of $\mathbf{\Phi}$, the factor variances, are allowed to become free parameters, though this would be unlikely to occur in practice.

7.7 STANDARD ERRORS OF ESTIMATES AND THE EFFECT OF STANDARDISATION

To find the standard errors of the estimates in $\hat{\boldsymbol{\theta}}$ for large samples the matrix $\mathbf{G}$ of section 7.2 is evaluated with $\boldsymbol{\theta} = \hat{\boldsymbol{\theta}}$ and inverted to give $\mathbf{E}$. The standard errors are then estimated as the square roots of the diagonal elements of the matrix $(2/n)\mathbf{E}$. This method of evaluation is correct, however, only if the covariance matrix $\mathbf{S}$ is unstandardised, and in practice it is generally more convenient to replace $\mathbf{S}$ by the sample correlation matrix having unit diagonal elements. We therefore consider what effect this has on the standard errors of estimates of the free parameters.

Let us henceforth assume, as is usually the case in practice, that all the ψ_i are free parameters and that all fixed parameters in $\mathbf{\Lambda}$ are zeros. It is then easy to verify that the method of estimation of this chapter is invariant under changes of scale of the x-variates. As in Chapter 5, let $\hat{\psi}_i^*$ denote the maximum likelihood estimate of ψ_i when the unstandardised covariance matrix $\mathbf{S}$ is replaced by the correlation matrix. Similarly let $\hat{\lambda}_{ir}^*$, be the corresponding maximum likelihood estimate of λ_{ir}, assuming that this is a free parameter. Then, as before,

$$\hat{\lambda}_{ir}^* = \hat{\lambda}_{ir}/\sqrt{s_{ii}},$$
$$\hat{\psi}_i^* = \hat{\psi}_i/s_{ii}.$$

Maximum likelihood estimates of any free parameters in $\mathbf{\Phi}$ are unaffected by standardisation.

Sampling formulae involving the $\hat{\lambda}_{ir}^*$ and the $\hat{\psi}_i^*$ may be found in much the same way as in section 5.3. With the notation of sections 7.2 and 7.3 and with $\boldsymbol{\theta}$ and $\partial F/\partial\boldsymbol{\theta}$ as there defined, we first consider the elements of $\partial F/\partial\boldsymbol{\theta}$ and we find the covariances of each of them with the diagonal elements of $\mathbf{S}$. This is accomplished by employing the methods of section 5.1. Assuming that λ_{ir} and

ϕ_{rs} are free parameters, we have, using (7.4), (7.5) and (7.6),

$$-n \operatorname{cov}(\partial F/\partial \lambda_{ir}, s_{jj}) = 4\delta_{ij}(\mathbf{\Lambda\Phi})_{jr},$$
$$-n \operatorname{cov}(\partial F/\partial \phi_{rs}, s_{jj}) = 4\lambda_{jr}\lambda_{js},$$
$$-n \operatorname{cov}(\partial F/\partial \psi_{i}, s_{jj}) = 2\delta_{ij}.$$

With the q elements of $\boldsymbol{\theta}$ and of $\partial F/\partial\boldsymbol{\theta}$ ordered as before and with the diagonal elements of $\mathbf{S}$ arranged in the order $s_{11}, \ldots, s_{pp}$, the above expressions form a $q \times p$ matrix $\mathbf{H}$. This is easily evaluated since most of its elements are zeros while the rest are very simple in form.

Asymptotically we have

$$\partial F/\partial\boldsymbol{\theta} = -\mathbf{G}(\hat{\boldsymbol{\theta}}-\boldsymbol{\theta})$$

and

$$\hat{\boldsymbol{\theta}}-\boldsymbol{\theta} = -\mathbf{G}^{-1}(\partial F/\partial\boldsymbol{\theta}) = -\mathbf{E}(\partial F/\partial\boldsymbol{\theta}).$$

Hence for large n, the covariance of the lth element of $\hat{\boldsymbol{\theta}}$ and s_{jj} is the element in the lth row and jth column of the $q \times p$ matrix $(1/n)\mathbf{EH}$.

As in (5.48) we have

$$\operatorname{cov}(\hat{\lambda}^*_{ir}, \hat{\lambda}^*_{jm}) = \operatorname{cov}(\hat{\lambda}_{ir}, \hat{\lambda}_{jm}) - \tfrac{1}{2}\lambda_{jm}\operatorname{cov}(\hat{\lambda}_{ir}, s_{jj}) - \tfrac{1}{2}\lambda_{ir}\operatorname{cov}(\hat{\lambda}_{jm}, s_{ii}) + \tfrac{1}{2}\lambda_{ir}\lambda_{jm}\sigma^2_{ij}/n. \qquad (7.16)$$

Similarly we have

$$\operatorname{cov}(\hat{\psi}^*_i, \hat{\psi}^*_j) = \operatorname{cov}(\hat{\psi}_i, \hat{\psi}_j) - \psi_j \operatorname{cov}(\hat{\psi}_i, s_{jj}) - \psi_i \operatorname{cov}(\hat{\psi}_j, s_{ii}) + 2\psi_i\psi_j\sigma^2_{ij}/n \qquad (7.17)$$

and

$$\operatorname{cov}(\hat{\lambda}^*_{ir}, \hat{\psi}^*_j) = \operatorname{cov}(\hat{\lambda}_{ir}, \hat{\psi}_j) - \psi_j \operatorname{cov}(\hat{\lambda}_{ir}, s_{jj}) - \tfrac{1}{2}\lambda_{ir}\operatorname{cov}(\hat{\psi}_j, s_{ii}) + \lambda_{ir}\psi_j\sigma^2_{ij}/n. \qquad (7.18)$$

We find also that

$$\operatorname{cov}(\hat{\lambda}^*_{ir}, \hat{\phi}_{mt}) = \operatorname{cov}(\hat{\lambda}_{ir}, \hat{\phi}_{mt}) - \tfrac{1}{2}\lambda_{ir}\operatorname{cov}(\hat{\phi}_{mt}, s_{ii})$$

and

$$\operatorname{cov}(\psi^*_i, \hat{\phi}_{rm}) = \operatorname{cov}(\hat{\psi}_i, \hat{\phi}_{rm}) - \psi_i \operatorname{cov}(\hat{\phi}_{rm}, s_{ii}).$$

As particular cases of (7.16) and (7.17) we have

$$\operatorname{var}(\hat{\lambda}^*_{ir}) = \operatorname{var}(\hat{\lambda}_{ir}) - \lambda_{ir}\operatorname{cov}(\hat{\lambda}_{ir}, s_{ii}) + \tfrac{1}{2}\lambda^2_{ir}\sigma^2_{ii}/n \qquad (7.19)$$

and

$$\operatorname{var}(\hat{\psi}^*_i) = \operatorname{var}(\hat{\psi}_i) - 2\psi_i\operatorname{cov}(\hat{\psi}_i, s_{ii}) + 2\psi^2_i\sigma^2_{ii}/n. \qquad (7.20)$$

In evaluating all the above expressions, $\hat{\lambda}^*_{ir}$, $\hat{\phi}_{rm}$ and $\hat{\psi}^*_i$ are substituted for λ_{ir}, ϕ_{rm} and ψ_i respectively whenever these are free parameters. For all values of i and j the σ_{ij} are found as elements of the matrix $\mathbf{\Lambda\Phi\Lambda'}+\mathbf{\Psi}$.

As a numerical example we consider the estimates of *Table 7.9*. The matrices **E** and **H** were evaluated and this enabled standard errors to be found both ignoring and allowing for the effect of standardisation. The results for the loadings and the residual variances are as shown in *Table 7.10*. Unbracketed entries represent standard errors corrected for standardisation, while those in brackets are the uncorrected standard errors. The factor correlation coefficients with standard errors attached are:

$$\hat{\phi}_{12} = 0{\cdot}55 \pm 0{\cdot}11,$$

$$\hat{\phi}_{13} = 0{\cdot}47 \pm 0{\cdot}14,$$

$$\hat{\phi}_{23} = 0{\cdot}09 \pm 0{\cdot}13.$$

Table 7.10 STANDARD ERRORS FOR ESTIMATES OF *Table 7.9*

Variate	$\hat{\boldsymbol{\lambda}}_1$	$\hat{\boldsymbol{\lambda}}_2$	$\hat{\boldsymbol{\lambda}}_3$	$\hat{\boldsymbol{\Psi}}$
1	0·09(0·12)	0	0	0·12(0·12)
2	0·12(0·13)	0	0	0·08(0·15)
3	0·09(0·12)	0	0	0·12(0·12)
4	0	0·04(0·10)	0	0·06(0·06)
5	0	0·04(0·10)	0	0·07(0·06)
6	0	0·05(0·10)	0	0·08(0·07)
7	0	0	0·10(0·13)	0·13(0·14)
8	0	0	0·11(0·14)	0·21(0·20)
9	0·11(0·13)	0	0·13(0·13)	0·11(0·10)

It is noticeable that correction for the effect of standardisation reduces the standard errors of the larger estimates in *Table 7.9*, in some cases considerably, whereas those of the smaller estimates are little affected. All estimates differ significantly from zero at the 5 per cent level with the exception of those for λ_{93}, ψ_8 and ϕ_{23}. It is interesting to note that when the RMLFA program was re-run with λ_{93} and ϕ_{23} both set equal to zero, and with 20 free parameters, an improper solution was obtained in which ψ_8 attained the boundary value of 0·005.

In view of its complexity, the results of this section (and, for that matter, those of Chapter 5) take the sampling theory as far as can reasonably be expected at present. However, they have much the same limitations as the well-known formula for the standard error of a sample correlation coefficient. For samples of the sizes commonly encountered in practice we are probably justified in assuming that maximum likelihood estimators of elements of $\mathbf{\Lambda}, \mathbf{\Phi}$ and $\mathbf{\Psi}$ are approximately normally distributed provided that, with standardised x-variates, these parameters do not exceed 0·6 in absolute value. For such a parameter we can find an approximate 95 per cent confidence interval by taking a range of twice the standard error on each side of its estimate. For a parameter whose absolute value is much greater than 0·6 such a procedure is hardly justifiable since the distribution of its estimator may well be far from normal.

Exercises

7.1. Let $(\mathbf{\Lambda}, \mathbf{\Phi}, \mathbf{\Psi})$ denote a unique maximum likelihood solution for which, in the notation of section 7.1, $n_\psi = 0$ and $n_\lambda + n_\phi = k^2$, so that there are altogether $pk - \frac{1}{2}k(k-1)$ free parameters in $\mathbf{\Lambda}$ and $\mathbf{\Phi}$. Show that the vanishing of elements of $\partial F/\partial \mathbf{\Lambda}$ and $\partial F/\partial \mathbf{\Phi}$ corresponding in position to these free parameters and also the symmetry of the matrix $\mathbf{\Lambda}'[\mathbf{\Sigma}^{-1}(\mathbf{\Sigma}-\mathbf{S})\mathbf{\Sigma}^{-1}\mathbf{\Lambda}]$ implies a set of pk linear, homogeneous and linearly independent equations for the pk elements of $\mathbf{\Sigma}^{-1}(\mathbf{\Sigma}-\mathbf{S})\mathbf{\Sigma}^{-1}\mathbf{\Lambda}$, thus showing that this matrix is null.

Let $\mathbf{M}$ be a $k \times k$ matrix such that $\mathbf{M\Phi M}' = \mathbf{I}$ and also such that $\mathbf{M}'^{-1}(\mathbf{\Lambda}'\mathbf{\Psi}^{-1}\mathbf{\Lambda})\mathbf{M}^{-1}$ is diagonal with its diagonal elements arranged in decreasing order of magnitude. Defining $\mathbf{\Lambda}^*$ by $\mathbf{\Lambda}^*\mathbf{M} = \mathbf{\Lambda}$, show using the result of the preceding paragraph, that $(\mathbf{\Lambda}^*, \mathbf{\Psi})$ is an unrestricted maximum likelihood solution satisfying the conditions and equations of estimation of section 4.2 of Chapter 4. As $\mathbf{\Lambda\Phi\Lambda}' = \mathbf{\Lambda}^*\mathbf{\Lambda}^{*\prime}$, this means that $(\mathbf{\Lambda}, \mathbf{\Phi}, \mathbf{\Psi})$ is not a genuinely restricted solution, since it can be obtained as a transformation of the orthogonal unrestricted solution $(\mathbf{\Lambda}^*, \mathbf{\Psi})$, and the two solutions have the same common factor space.

Verify that the number of degrees of freedom given in (7.14) for the χ^2 test of (7.15) agrees in this case with that for the test of (4.30).

7.2. Suppose that all fixed parameters in $\mathbf{\Lambda}$ are zeros and that all the ψ_i and all non-diagonal elements of $\mathbf{\Phi}$ are free parameters. Dropping circumflex accents, let $\mathbf{\Lambda}, \mathbf{\Phi}, \mathbf{\Psi}$ and $\mathbf{\Sigma}$ refer to maximum likelihood estimates of parameters. Then the elements of the matrices on the right-hand sides of (7.4), (7.5) and (7.6) that correspond in position to free parameters in $\mathbf{\Lambda}$, $\mathbf{\Phi}$ and $\mathbf{\Psi}$ respectively may be equated to zero.

Show that

$$\text{diag}\,[\mathbf{\Lambda}'\mathbf{\Sigma}^{-1}(\mathbf{\Sigma}-\mathbf{S})\mathbf{\Sigma}^{-1}\mathbf{\Lambda\Phi}] = \mathbf{0}$$

and hence that the matrix on the right-hand side of (7.5) is null. Deduce from this that

$$\mathbf{\Lambda}'\mathbf{\Psi}^{-1}(\mathbf{\Sigma}-\mathbf{S})\mathbf{\Psi}^{-1}\mathbf{\Lambda} = \mathbf{0}$$

and hence that $\mathbf{\Phi}$ is given in terms of $\mathbf{\Lambda}$, $\mathbf{\Psi}$ and $\mathbf{S}$ by

$$\mathbf{\Phi} = \mathbf{\Gamma}^{-1}(\mathbf{\Lambda}'\mathbf{\Psi}^{-1}\mathbf{S}\mathbf{\Psi}^{-1}\mathbf{\Lambda})\mathbf{\Gamma}^{-1}-\mathbf{\Gamma}^{-1},$$

where

$$\mathbf{\Gamma} = \mathbf{\Lambda}'\mathbf{\Psi}^{-1}\mathbf{\Lambda}.$$

The last result enables the minimisation of F in this case to be carried out with respect to $\mathbf{\Lambda}$ and $\mathbf{\Psi}$ alone.

7.3. With the same assumptions and notation as in Exercise 7.2, pre-multiply the matrix on the right-hand side of (7.4) by $\mathbf{\Psi}$ in the form $\mathbf{\Sigma}-\mathbf{\Lambda\Phi\Lambda}'$ and use the fact that the matrix on the right-hand side of (7.5) is null. Hence show that the elements of the matrix $(\mathbf{\Sigma}-\mathbf{S})\mathbf{\Sigma}^{-1}\mathbf{\Lambda\Phi}$ corresponding in position to free parameters in $\mathbf{\Lambda}$ are zero. Deduce that

$$\text{diag}\,[(\mathbf{\Sigma}-\mathbf{S})\mathbf{\Sigma}^{-1}\mathbf{\Lambda\Phi\Lambda}'] = \mathbf{0}.$$

Now post-multiply the matrix on the right-hand side of (7.6) by $\mathbf{\Psi}$ in the form $\mathbf{\Sigma}-\mathbf{\Lambda\Phi\Lambda}'$ and use the last result to show that

$$\text{diag}\,[\mathbf{\Sigma}^{-1}(\mathbf{\Sigma}-\mathbf{S})] = \mathbf{0}.$$

Finally, pre-multiply this equation by $\mathbf{\Psi}$ in the same form as before and show that

$$\text{diag}\,(\mathbf{\Sigma}) = \text{diag}\,(\mathbf{S}).$$

Thus the diagonal elements of $\mathbf{S}$ are, in this case, the maximum likelihood estimates of the variances.

7.4. Maximum likelihood factor analyses were carried out on the data of *Table 6.1* under two different hypotheses, as follows:

(a) with two correlated factors, the parameters λ_{41}, λ_{51}, λ_{61}, λ_{12}, λ_{22} and λ_{32} were all set equal to zero. The remaining loadings,

the residual variances and the factor correlation coefficient ϕ_{12} were all free parameters. The loadings were estimated as:

0·687	0·672	0·533	0	0	0
0	0	0	0·766	0·768	0·616

and the residual variances as:

0·528	0·548	0·716	0·413	0·410	0·621

The estimate of ϕ_{12} was 0·597.

(b) with two orthogonal factors, the parameters λ_{42}, λ_{52}, λ_{62} and ϕ_{12} were all set equal to zero. The remaining loadings and the residual variances were all free parameters. The loadings were estimated as:

0·402	0·456	0·242	0·771	0·763	0·617
0·576	0·443	0·543	0	0	0

and the residual variances as:

0·506	0·595	0·647	0·406	0·417	0·619

For analysis *(a)* the minimum function value $F(\hat{\theta})$ was 0·036316, while for analysis *(b)* it was 0·014764. Show that in each case the test of (7.15) leads to acceptance of the hypothesis, the value of χ^2 being below expectation.

In case *(a)* verify the last result of Exercise 7.2.

Chapter 8

THE ESTIMATION OF FACTOR SCORES

8.1 INTRODUCTION

So far we have been interested mainly in problems concerning the parameters in factor models and their estimation. While these problems constitute the chief interest of factor analysis, it is sometimes desirable to go a step further and to estimate the scores of an individual on the hypothetical factors in terms of the observations of the x-variates for that individual. We therefore turn our attention to the problem of estimating factor scores.

8.2 THE REGRESSION METHOD WITH UNCORRELATED FACTORS

We shall first suppose that the factors are uncorrelated and standardised. The usual distributional assumptions are made regarding the factor scores f_r and the residuals e_i. We assume that the true values of all parameters are known. In practice $\mathbf{\Lambda}$ and $\mathbf{\Psi}$ would usually have been estimated by some method, but in such cases it is supposed that sampling errors may be ignored.

It is clear that factor scores cannot be estimated in the usual statistical sense since they are not parameters but values ascribed to unobservable variates. Furthermore, the unobservables include not only the k factor scores $f_1, \ldots, f_k$ but also the p residuals $e_1, \ldots, e_p$. They thus outnumber the observations $x_1, \ldots, x_p$. In order to provide reasonable estimates of the factor scores it is necessary to introduce some new principle of a minimum variance or least squares type.

Let $\mathbf{x}$ be the vector of observations $x_1, \ldots, x_p$. Let $\mathbf{f}$ be the vector of factor scores and $\mathbf{e}$ the vector of residuals. Then we have, as formerly,

$$E(\mathbf{x}\mathbf{f}') = E[(\mathbf{\Lambda}\mathbf{f}+\mathbf{e})\mathbf{f}'] = \mathbf{\Lambda}E(\mathbf{f}\mathbf{f}') = \mathbf{\Lambda}$$

and

$$E(\mathbf{xx}') = \mathbf{\Sigma} = \mathbf{\Lambda\Lambda}'+\mathbf{\Psi}.$$

For $r = 1, \ldots, k$ we seek a linear function of the observations that will provide a good predictor of f_r. Let this be

$$\hat{f}_r = \mathbf{a}_r'\mathbf{x} = \mathbf{x}'\mathbf{a}_r,$$

where $\mathbf{a}_r$ is a vector of order p. We shall choose $\mathbf{a}_r$ in such a way that the variance of $\hat{\mathbf{f}}_r - \mathbf{f}_r$ is minimised. This variance is

$$E(\hat{f}_r - f_r)^2 = E(\mathbf{x}'\mathbf{a}_r - f_r)^2.$$

To minimise this expression we equate to zero its derivative with respect to the vector $\mathbf{a}_r$. The derivative is

$$E[2\mathbf{x}(\mathbf{x}'\mathbf{a}_r - f_r)] = 2(\mathbf{\Sigma a}_r - \boldsymbol{\lambda}_r),$$

where $\boldsymbol{\lambda}_r$ is the rth column of $\mathbf{\Lambda}$. Hence we have

$$\mathbf{\Sigma a}_r = \boldsymbol{\lambda}_r,$$

or

$$\mathbf{a}_r = \mathbf{\Sigma}^{-1}\boldsymbol{\lambda}_r,$$

and $\hat{f}_r$ is given by

$$\hat{f}_r = \boldsymbol{\lambda}_r'\mathbf{\Sigma}^{-1}\mathbf{x}.$$

If $\hat{\mathbf{f}}$ denotes the vector of estimates $\hat{f}_1, \ldots, \hat{f}_k$, we may write

$$\hat{\mathbf{f}} = \mathbf{\Lambda}'\mathbf{\Sigma}^{-1}\mathbf{x}. \tag{8.1}$$

The above method of obtaining $\hat{\mathbf{f}}$ is equivalent to finding the linear regression relationship of $\mathbf{f}$ on $\mathbf{x}$. Hence we shall in future term it the regression method. A discussion of it has been given by Thomson (1951), by whom the estimates of (8.1) were first derived.

By use of the usual identity for $\mathbf{\Lambda}'\mathbf{\Sigma}^{-1}$, equation (8.1) may be put in the alternative form

$$\hat{\mathbf{f}} = (\mathbf{I}+\mathbf{\Gamma})^{-1}\mathbf{\Lambda}'\mathbf{\Psi}^{-1}\mathbf{x}, \tag{8.2}$$

where

$$\mathbf{\Gamma} = \mathbf{\Lambda}'\mathbf{\Psi}^{-1}\mathbf{\Lambda}.$$

The covariance matrix for the estimates is $E(\hat{\mathbf{f}}\hat{\mathbf{f}}')$. The covariances between true and estimated factors are the elements of the matrix $E(\hat{\mathbf{f}}\mathbf{f}')$. These matrices are given by

$$\begin{aligned} E(\hat{\mathbf{f}}\hat{\mathbf{f}}') = E(\hat{\mathbf{f}}\mathbf{f}') &= \mathbf{\Lambda}'\mathbf{\Sigma}^{-1}\mathbf{\Lambda} \\ &= \mathbf{I}-(\mathbf{I}+\mathbf{\Gamma})^{-1}. \end{aligned} \tag{8.3}$$

For the errors of prediction, which are the elements of $\hat{\mathbf{f}}-\mathbf{f}$, the covariance matrix is

$$E[(\hat{\mathbf{f}}-\mathbf{f})(\hat{\mathbf{f}}-\mathbf{f})'] = \mathbf{I}-\boldsymbol{\Lambda}'\boldsymbol{\Sigma}^{-1}\boldsymbol{\Lambda}$$
$$= (\mathbf{I}+\boldsymbol{\Gamma})^{-1}. \tag{8.4}$$

It may also be verified that the rth diagonal element of the matrix in (8.3) is the square of the correlation coefficient between $\hat{f}_r$ and f_r.

From the above results it is clear that for good prediction we require the diagonal elements of $(\mathbf{I}+\boldsymbol{\Gamma})^{-1}$ to be small. If the factors are defined in such a way that $\boldsymbol{\Gamma}$ is diagonal, it is necessary that the diagonal elements of $\boldsymbol{\Gamma}$ should be large. This is usually true of the first few elements, when they are arranged in decreasing order of magnitude. Hence the first few factors would, as a rule, be predicted reasonably well; but the prediction tends to be far less good for the later factors.

As an example consider data 1 of Chapter 4, section 4.6. In the notation of that chapter, $\mathbf{I}+\boldsymbol{\Gamma}$ is the diagonal matrix $\boldsymbol{\Theta}$, whose diagonal elements are estimated as 15·968, 4·358 and 1·848. If we neglect errors of estimation, the matrix in (8.3) is diagonal with elements 0·937, 0·771 and 0·459. The correlation coefficients between $\hat{f}_r$ and f_r for $r = 1, 2, 3$ are the square roots of these values and are respectively 0·968, 0·878 and 0·677. Evidently f_1, and to a lesser extent f_2, can be well estimated; but this is not true of f_3.

We now show that $\hat{\mathbf{f}}$, like other regression estimates, is in a sense biased, since its expectation for given $\mathbf{f}$ is not equal to $\mathbf{f}$. In other words, if we average $\hat{\mathbf{f}}$ over all individuals whose true factor scores are given by $\mathbf{f}$, the result differs from $\mathbf{f}$. Since $\mathbf{e}$ is independent of $\mathbf{f}$, the expectation of $\mathbf{x}$ for given $\mathbf{f}$ is

$$E(\mathbf{x}\,|\,\mathbf{f}) = E[(\boldsymbol{\Lambda}\mathbf{f}+\mathbf{e})\,|\,\mathbf{f}] = \boldsymbol{\Lambda}\mathbf{f}.$$

Hence

$$E(\hat{\mathbf{f}}\,|\,\mathbf{f}) = E(\boldsymbol{\Lambda}'\boldsymbol{\Sigma}^{-1}\mathbf{x}\,|\,\mathbf{f})$$
$$= (\boldsymbol{\Lambda}'\boldsymbol{\Sigma}^{-1}\boldsymbol{\Lambda})\mathbf{f}$$
$$= \mathbf{f}-(\mathbf{I}+\boldsymbol{\Gamma})^{-1}\mathbf{f}. \tag{8.5}$$

In section 8.4 we find an estimate of $\mathbf{f}$ that is not biased in this way.

8.3 THE REGRESSION METHOD WITH CORRELATED FACTORS

The regression method of estimating $\mathbf{f}$ may without difficulty be generalised to cases where the factors are correlated and not necessarily standardised. Denoting the factor covariance matrix, as before, by $\mathbf{\Phi}$, we now have

$$E(\mathbf{ff}') = \mathbf{\Phi},$$
$$E(\mathbf{xf}') = E[(\mathbf{\Lambda f}+\mathbf{e})\mathbf{f}'] = \mathbf{\Lambda\Phi},$$
$$E(\mathbf{xx}') = \mathbf{\Sigma} = \mathbf{\Lambda\Phi\Lambda}'+\mathbf{\Psi}.$$

The algebra proceeds in the same way as before. For $\hat{\mathbf{f}}$ we now have

$$\hat{\mathbf{f}} = \mathbf{\Phi\Lambda}'\mathbf{\Sigma}^{-1}\mathbf{x}, \tag{8.6}$$

or

$$\hat{\mathbf{f}} = \mathbf{\Phi}(\mathbf{I}+\mathbf{\Gamma\Phi})^{-1}\mathbf{\Lambda}'\mathbf{\Psi}^{-1}\mathbf{x}, \tag{8.7}$$

where $\mathbf{\Gamma}$ is as formerly defined. In place of equations (8.3) to (8.5) we have

$$E(\hat{\mathbf{f}}\hat{\mathbf{f}}') = E(\hat{\mathbf{f}}\mathbf{f}') = \mathbf{\Phi\Lambda}'\mathbf{\Sigma}^{-1}\mathbf{\Lambda\Phi}$$
$$= \mathbf{\Phi}-\mathbf{\Phi}(\mathbf{I}+\mathbf{\Gamma\Phi})^{-1}, \tag{8.8}$$

$$E[(\hat{\mathbf{f}}-\mathbf{f})(\hat{\mathbf{f}}-\mathbf{f})'] = \mathbf{\Phi}(\mathbf{I}+\mathbf{\Gamma\Phi})^{-1} \tag{8.9}$$

and

$$E(\hat{\mathbf{f}}\,|\,\mathbf{f}) = \mathbf{\Phi}(\mathbf{\Lambda}'\mathbf{\Sigma}^{-1}\mathbf{\Lambda})\mathbf{f}$$
$$= \mathbf{\Phi}(\mathbf{I}+\mathbf{\Gamma\Phi})^{-1}(\mathbf{\Lambda}'\mathbf{\Psi}^{-1}\mathbf{\Lambda})\mathbf{f}$$
$$= (\mathbf{I}+\mathbf{\Phi\Gamma})^{-1}\mathbf{\Phi\Gamma f}$$
$$= \mathbf{f}-(\mathbf{I}+\mathbf{\Phi\Gamma})^{-1}\mathbf{f}. \tag{8.10}$$

The above results reduce to those of section 8.2 on putting $\mathbf{\Phi} = \mathbf{I}$.

8.4 AN ALTERNATIVE METHOD OF ESTIMATION

A different method of estimating factor scores was proposed by Bartlett (1937, 1938). His estimates may be derived by adopting a least squares principle and minimising the sum of squares of the standardised residuals, which is

$$\sum_{i=1}^{p}(e_i^2/\psi_i) = \mathbf{e}'\mathbf{\Psi}^{-1}\mathbf{e} = (\mathbf{x}-\mathbf{\Lambda f})'\mathbf{\Psi}^{-1}(\mathbf{x}-\mathbf{\Lambda f}).$$

The minimisation is with respect to the elements of $\mathbf{f}$, regarded now as mathematical variables. To minimise the above expression we equate to zero its derivative with respect to the vector $\mathbf{f}$, which is

$$-2\mathbf{\Lambda}'\mathbf{\Psi}^{-1}(\mathbf{x}-\mathbf{\Lambda f}) = 2(\mathbf{\Gamma f}-\mathbf{\Lambda}'\mathbf{\Psi}^{-1}\mathbf{x}).$$

Denoting the estimate that satisfies this equation by $\hat{\mathbf{f}}^*$, we have

$$\mathbf{\Gamma}\hat{\mathbf{f}}^* = \mathbf{\Lambda}'\mathbf{\Psi}^{-1}\mathbf{x},$$

or

$$\hat{\mathbf{f}}^* = \mathbf{\Gamma}^{-1}\mathbf{\Lambda}'\mathbf{\Psi}^{-1}\mathbf{x}. \tag{8.11}$$

The form of this estimate does not depend upon whether the factors are correlated or not.

On comparing (8.11) with (8.7) we find that there is a simple relationship between $\hat{\mathbf{f}}^*$ and $\hat{\mathbf{f}}$. We have

$$\hat{\mathbf{f}}^* = (\mathbf{I}+\mathbf{\Gamma}^{-1}\mathbf{\Phi}^{-1})\hat{\mathbf{f}}, \tag{8.12}$$

or

$$\hat{\mathbf{f}} = \mathbf{\Phi}\mathbf{\Gamma}(\mathbf{I}+\mathbf{\Phi}\mathbf{\Gamma})^{-1}\hat{\mathbf{f}}^*. \tag{8.13}$$

In the case where the factors are uncorrelated and where $\mathbf{\Gamma}$ is diagonal the estimate of f_r in $\hat{\mathbf{f}}^*$ differs from that in $\hat{\mathbf{f}}$ only by a scale factor.

The various covariance matrices involving $\hat{\mathbf{f}}^*$ are

$$E(\hat{\mathbf{f}}^*\mathbf{f}') = \mathbf{\Phi},$$

$$E(\hat{\mathbf{f}}^*\hat{\mathbf{f}}^{*\prime}) = \mathbf{\Phi}+\mathbf{\Gamma}^{-1}$$

and

$$E[(\hat{\mathbf{f}}^*-\mathbf{f})(\hat{\mathbf{f}}^*-\mathbf{f})'] = \mathbf{\Gamma}^{-1}.$$

We have also

$$E(\hat{\mathbf{f}}^*\,|\,\mathbf{f}) = E(\mathbf{\Gamma}^{-1}\mathbf{\Lambda}'\mathbf{\Psi}^{-1}\mathbf{x}\,|\,\mathbf{f}) = \mathbf{\Gamma}^{-1}\mathbf{\Lambda}'\mathbf{\Psi}^{-1}\mathbf{\Lambda}\mathbf{f} = \mathbf{f}.$$

Thus $\hat{\mathbf{f}}^*$ is unbiased in the same sense as before.

Bartlett's estimates may be derived in a different way that is more in accord with our previous approach. Let us restrict ourselves to the class of unbiased estimates and, for $r = 1, \ldots, k$, find the unbiased estimate $\hat{f}_r$ for which the variance of $\hat{f}_r - f_r$ is minimised. As we shall show, the resulting estimates are the elements of $\hat{\mathbf{f}}^*$.

Let $\hat{f}_r = \mathbf{a}_r'\mathbf{x}$ be the required unbiased estimate of f_r. Then the vector $\hat{\mathbf{f}}$ is given by

$$\hat{\mathbf{f}} = \mathbf{A}'\mathbf{x},$$

where $\mathbf{A}$ is the $p \times k$ matrix whose rth column is $\mathbf{a}_r$. Since

$$E(\hat{\mathbf{f}}\,|\,\mathbf{f}) = \mathbf{A}'\mathbf{\Lambda}\mathbf{f},$$

the condition that $\hat{\mathbf{f}}$ should be unbiased is that

$$\mathbf{A}'\mathbf{\Lambda} = \mathbf{\Lambda}'\mathbf{A} = \mathbf{I}. \tag{8.14}$$

This is equivalent to

$$\mathbf{a}_r'\mathbf{\Lambda} = \boldsymbol{\delta}_r' \qquad (r = 1, \ldots, k),$$

where $\boldsymbol{\delta}_r$ is the vector of order k whose rth element is unity and whose other elements are zeros.

We now choose $\mathbf{a}_r$, subject to the above condition, such that $E(\mathbf{x}'\mathbf{a}_r - f_r)^2$, the variance of $\hat{f}_r - f_r$, is minimised. We do this by equating to zero the derivative with respect to $\mathbf{a}_r$ of the expression

$$E[(\mathbf{x}'\mathbf{a}_r - f_r)^2] - 2\mathbf{a}_r'\mathbf{\Lambda}\mathbf{c}_r,$$

where $\mathbf{c}_r$ is a vector of order k whose elements are indeterminate multipliers. The derivative is

$$2E[\mathbf{x}(\mathbf{x}'\mathbf{a}_r - f_r)] - 2\mathbf{\Lambda}\mathbf{c}_r = 2(\mathbf{\Sigma}\mathbf{a}_r - \boldsymbol{\lambda}_r - \mathbf{\Lambda}\mathbf{c}_r),$$

where $\boldsymbol{\lambda}_r$ is, as before, the rth column of $\mathbf{\Lambda}$. On equating this to zero we have

$$\mathbf{\Sigma}\mathbf{a}_r = \boldsymbol{\lambda}_r + \mathbf{\Lambda}\mathbf{c}_r.$$

This equation, when taken for all values of r, is equivalent to

$$\mathbf{\Sigma}\mathbf{A} = \mathbf{\Lambda}(\mathbf{I} + \mathbf{C}),$$

where $\mathbf{C}$ is the $k \times k$ matrix whose rth column is $\mathbf{c}_r$.

Hence

$$\begin{aligned}\mathbf{A} &= \mathbf{\Sigma}^{-1}\mathbf{\Lambda}(\mathbf{I} + \mathbf{C}) \\ &= \mathbf{\Psi}^{-1}\mathbf{\Lambda}(\mathbf{I} + \mathbf{\Phi}\mathbf{\Gamma})^{-1}(\mathbf{I} + \mathbf{C}).\end{aligned}$$

Let us write this more simply as

$$\mathbf{A} = \mathbf{\Psi}^{-1}\mathbf{\Lambda}\mathbf{G}.$$

To determine $\mathbf{G}$ we use the condition (8.14) for lack of bias, which gives

$$\mathbf{\Lambda}'\mathbf{\Psi}^{-1}\mathbf{\Lambda}\mathbf{G} = \mathbf{I}.$$

Hence

$$\mathbf{G} = (\mathbf{\Lambda}'\mathbf{\Psi}^{-1}\mathbf{\Lambda})^{-1} = \mathbf{\Gamma}^{-1}$$

and

$$\mathbf{A} = \mathbf{\Psi}^{-1}\mathbf{\Lambda}\mathbf{\Gamma}^{-1}.$$

Thus $\hat{\mathbf{f}} = \mathbf{A}'\mathbf{x}$ is identical with $\hat{\mathbf{f}}^*$.

Of the two methods of estimating factor scores the one that should be adopted in any particular circumstances depends solely on the properties that the estimates are required to possess. No general preference can therefore be given. Bartlett's estimates achieve lack of bias, but at the cost of having greater variability than the regression estimates.

8.5 A NUMERICAL EXAMPLE

As an example of the estimation of factor scores (with uncorrelated factors) we take $\mathbf{\Lambda}$ to be the matrix of loadings given in *Table 6.2*. The elements of $\mathbf{\Psi}$ are obtained by subtracting the communalities from unity and are thus:

$$0{\cdot}510 \quad 0{\cdot}594 \quad 0{\cdot}644 \quad 0{\cdot}377 \quad 0{\cdot}431 \quad 0{\cdot}628$$

The matrix $\mathbf{\Gamma}$ is diagonal, and its diagonal elements are 4·614 and 1·143.

Using these results we find that the elements of $(\mathbf{I}+\mathbf{\Gamma})^{-1}\mathbf{\Lambda}'\mathbf{\Psi}^{-1}$ are as follows:

0·193	0·170	0·108	0·350	0·299	0·169
0·392	0·226	0·326	−0·338	−0·229	−0·098

These coefficients enable us to find the regression estimates of f_1 and f_2. Thus, for example,

$$\hat{f}_1 = 0{\cdot}193x_1+0{\cdot}170x_2+\ldots+0{\cdot}169x_6,$$

where the x's are standardised observations. The estimation of f_1 is reasonably good since the correlation coefficient between $\hat{f}_1$ and f_1 is $\sqrt{(4{\cdot}614/5{\cdot}614)} = 0{\cdot}907$. However, f_2 is not as well estimated since the corresponding correlation coefficient for it is only 0·730.

The elements of the matrix $\mathbf{\Gamma}^{-1}\mathbf{\Lambda}'\mathbf{\Psi}^{-1}$ are as follows:

0·235	0·207	0·132	0·425	0·364	0·205
0·736	0·424	0·612	−0·633	−0·429	−0·184

From these coefficients Bartlett's estimates of f_1 and f_2 may be found. There is clearly a large difference in scaling between $\hat{f}_2$ and $\hat{f}_2^*$.

Exercises

8.1. *Table 8.1* gives the estimated loadings on two correlated factors and the residual variances of the 10 sub-tests of the Wechsler Pre-School and Primary Scale of Intelligence (omitting the sub-test 'sentences'). The correlation coefficient between the two factors is 0·763.

Ignoring the sampling errors of these estimates, find the unbiased estimates $\hat{f}_1^*$ and $\hat{f}_2^*$ of the two factors f_1 and f_2. Show that the correlation coefficients between $\hat{f}_1^*$ and f_1 and between $\hat{f}_2^*$ and f_2 are 0·947 and 0·929 respectively.

8.2. Find the regression estimates $\hat{f}_1$ and $\hat{f}_2$ of the two factors for the data of Exercise 8.1. Show that the correlation coefficients between $\hat{f}_1$ and f_1 and between $\hat{f}_2$ and f_2 are 0·952 and 0·939 respectively.

Table 8.1

Sub-test	*Loadings* f_1	f_2	*Residual variance*
1	0·838	0	0·297
2	0·884	0	0·219
3	0·734	0	0·461
4	0·609	0	0·629
5	0·720	0	0·482
6	0	0·689	0·525
7	0	0·705	0·502
8	0	0·715	0·488
9	0	0·560	0·687
10	0	0·863	0·256

8.3. For the data of Exercise 8.1, find the values of a_1 and a_2 such that

$$y_1 = a_1(x_1+x_2+\ldots+x_5)$$

and

$$y_2 = a_2(x_6+x_7+\ldots+x_{10})$$

are, in the sense of section 8.4, unbiased estimates of f_1 and f_2 respectively.

Find the correlation coefficients between y_1 and f_1 and between y_2 and f_2. These show that y_1 and y_2 compare favourably with $\hat{f}_1^*$ and $\hat{f}_2^*$ as unbiased estimates of the two factors.

Chapter 9

IDENTIFYING FACTORS IN DIFFERENT POPULATIONS

9.1 INTRODUCTION

Many difficult problems in factor analysis arise when we wish to compare results obtained from different sources, and much research is still required in this field. Some of the problems originated quite early in the history of the subject and have been well summarised by Thomson (1951) and Thurstone (1947) in their respective textbooks.

Thomson was particularly interested in the effect produced on factors by selection of one or more of the variates. By employing the selection formulae first given by Pearson (1912), and later expressed in matrix form by Aitken (1934), he was able to show that such selection may cause factors that were originally orthogonal to become correlated. He also found that additional factors could be introduced having loadings in the variates directly selected. It follows that, if we wish to identify the factors that appear when two different populations are sampled, we shall not necessarily find it useful to analyse separately the covariance or correlation matrices obtained from the two samples, especially if the factors are in each case restricted to being orthogonal. When more than one common factor is involved it is often difficult to discover the relationship between the factors found in one sample and those found in the other.

As a result of his investigations Thomson came to somewhat pessimistic conclusions regarding the permanence of factors, even for a given set of variates. A more hopeful attitude was adopted by Thurstone, who went some way towards overcoming the difficulties by introducing his idea of 'simple structure'.

In an attempt to clarify and resolve some of the problems of identifying factors in different populations we have been led to propose a simple model, the basic assumption of which is that any selective process operates directly on the factors and only indirectly on the variates. Thus we consider multivariate populations in which the factors remain the same and which differ only as regards the variances and covariances of the factors.

9.2 A HYPOTHESIS FOR TWO POPULATIONS

Let us suppose that for each of two p-variate normal populations the same k factors are in operation. The covariance matrices of the factors in the two populations are denoted respectively by $\mathbf{\Phi}_1$, and $\mathbf{\Phi}_2$. The relationship between the variates x_i ($i = 1, \ldots, p$) and the factors f_r ($r = 1, \ldots, k$) is, as usual, given by equations (1.1) or, in matrix form, by equation (2.1). The coefficients λ_{ir} in these equations are invariant and hence the $p \times k$ loading matrix $\mathbf{\Lambda} = [\lambda_{ir}]$ is the same for both populations.

Partly for simplicity and partly because there may be practical justification, we shall assume that the residual variances of the x-variates are the same for both populations and that in each they form a diagonal matrix $\mathbf{\Psi}$. The covariance matrices of the x-variates in the two populations are then given respectively by

$$\mathbf{\Sigma}_1 = \mathbf{\Lambda}\mathbf{\Phi}_1\mathbf{\Lambda}' + \mathbf{\Psi}$$

and

$$\mathbf{\Sigma}_2 = \mathbf{\Lambda}\mathbf{\Phi}_2\mathbf{\Lambda}' + \mathbf{\Psi}. \tag{9.1}$$

As in Chapter 7 we shall suppose that certain elements of $\mathbf{\Lambda}$ are zero and that the number and positions of these loadings are such that the factors are uniquely determined. Since the scales of the factors are arbitrary we shall, purely for convenience, assume that they all have unit variances in the first population, but not necessarily in the second. Thus the diagonal elements of $\mathbf{\Phi}_1$ are fixed parameters having the value unity whereas those of $\mathbf{\Phi}_2$ are free parameters. For the sake of simplicity we assume that all non-diagonal elements of $\mathbf{\Phi}_1$ and $\mathbf{\Phi}_2$ and all residual variances are also free parameters. Thus if n_λ denotes the number of elements of $\mathbf{\Lambda}$ that are restricted to be zero, the total number of fixed parameters is $m = n_\lambda + k$, while the total number of free parameters is

$$\begin{aligned} q &= pk + k(k+1) + p - m \\ &= k(p+k) + p - n_\lambda. \end{aligned} \tag{9.2}$$

9.3 ESTIMATION OF PARAMETERS

With the usual distributional assumptions regarding the f_r and the e_i within each population, we suppose that random samples of respective sizes N_1 and N_2 are drawn from the two populations. Let $\mathbf{S}_1$ and $\mathbf{S}_2$ denote the usual sample estimators of $\boldsymbol{\Sigma}_1$ and $\boldsymbol{\Sigma}_2$ having respectively $n_1 = N_1 - 1$ and $n_2 = N_2 - 1$ degrees of freedom. We now consider how these may be used to provide estimates of all the free parameters. Before employing the maximum likelihood method we describe a simple procedure for obtaining a set of initial estimates.

Let $\mathbf{S}$ be the pooled 'within samples' covariance matrix defined by

$$\mathbf{S} = (1/n)\,(n_1\mathbf{S}_1 + n_2\mathbf{S}_2), \tag{9.3}$$

where $n = n_1 + n_2$. Then since $E(\mathbf{S}_1) = \boldsymbol{\Sigma}_1$ and $E(\mathbf{S}_2) = \boldsymbol{\Sigma}_2$, we have, using (9.1),

$$E(\mathbf{S}) = (1/n)\,(n_1\boldsymbol{\Sigma}_1 + n_2\boldsymbol{\Sigma}_2)$$
$$= \boldsymbol{\Lambda\Phi\Lambda}' + \boldsymbol{\Psi},$$

where

$$\boldsymbol{\Phi} = (1/n)\,(n_1\boldsymbol{\Phi}_1 + n_2\boldsymbol{\Phi}_2).$$

On analysing the matrix $\mathbf{S}$ by one of the methods described in Chapter 6, rough estimates of $\boldsymbol{\Lambda}$, $\boldsymbol{\Psi}$ and $\boldsymbol{\Phi}$ are obtained. The estimate of $\boldsymbol{\Phi}$ is not of interest; but let the other two estimates be denoted by $\boldsymbol{\Lambda}^*$ and $\hat{\boldsymbol{\Psi}}$. The asterisk is used here to indicate that the factors and the loadings on them will require a final re-scaling.

To estimate the covariance matrices of the factors in the two populations we may first find, for $h = 1, 2$,

$$\boldsymbol{\Phi}_h^* = \boldsymbol{\Gamma}^{-1}\boldsymbol{\Lambda}^{*\prime}\boldsymbol{\Psi}^{-1}(\mathbf{S}_h - \boldsymbol{\Psi})\boldsymbol{\Psi}^{-1}\boldsymbol{\Lambda}^*\boldsymbol{\Gamma}^{-1},$$

where

$$\boldsymbol{\Gamma} = \boldsymbol{\Lambda}^{*\prime}\boldsymbol{\Psi}^{-1}\boldsymbol{\Lambda}^*.$$

We then re-scale the factors so that each is estimated to have unit variance in the first population. For this purpose let $\mathbf{D}$ be the diagonal matrix with positive diagonal elements such that

$$\mathbf{D}^2 = \operatorname{diag}(\boldsymbol{\Phi}_1^*).$$

Then the loading matrix for the re-scaled factors is estimated as

$$\hat{\boldsymbol{\Lambda}} = \boldsymbol{\Lambda}^*\mathbf{D},$$

while the estimates of the factor covariance matrices in the two populations are given respectively by

$$\hat{\boldsymbol{\Phi}}_1 = \mathbf{D}^{-1}\boldsymbol{\Phi}_1^*\mathbf{D}^{-1}$$

and

$$\hat{\mathbf{\Phi}}_2 = \mathbf{D}^{-1}\mathbf{\Phi}_2^*\mathbf{D}^{-1}.$$

The first of these has unit diagonal elements, as required.

Since $n_1\mathbf{S}_1$ and $n_2\mathbf{S}_2$ follow independent Wishart distributions, being respectively $W(\mathbf{\Sigma}_1, n_1)$ and $W(\mathbf{\Sigma}_2, n_2)$, the log-likelihood function based on the information provided by $\mathbf{S}_1$ and $\mathbf{S}_2$ is, ignoring a function of the observations,

$$-\tfrac{1}{2}n_1[\log_e|\mathbf{\Sigma}_1|+\text{tr}\,(\mathbf{S}_1\mathbf{\Sigma}_1^{-1})]-\tfrac{1}{2}n_2[\log_e|\mathbf{\Sigma}_2|+\text{tr}\,(\mathbf{S}_2\mathbf{\Sigma}_2^{-1})]. \quad (9.4)$$

To obtain maximum likelihood estimates it is necessary to maximise the above expression with respect to all free parameters. In actual practice it would be more convenient to minimise the function $F = F(\mathbf{\Lambda}, \mathbf{\Phi}_1, \mathbf{\Phi}_2, \mathbf{\Psi})$ given by

$$n_1 \log|\mathbf{\Sigma}_1|+n_2 \log|\mathbf{\Sigma}_2|+n_1\,\text{tr}\,(\mathbf{S}_1\mathbf{\Sigma}_1^{-1})+n_2\,\text{tr}\,(\mathbf{S}_2\mathbf{\Sigma}_2^{-1})$$
$$-n_1 \log|\mathbf{S}_1|-n_2 \log|\mathbf{S}_2|-np. \quad (9.5)$$

This differs from -2 times the expression of (9.4) only by a function of the observations.

The function F can be minimised by an iterative procedure of the same kind as that employed in Chapter 7, using the estimates previously obtained as initial values. As before, the first-order derivatives of F with respect to the free parameters are required. We have

$$\partial F/\partial\mathbf{\Lambda} = 2n_1\mathbf{V}_1+2n_2\mathbf{V}_2, \quad (9.6)$$

where

$$\mathbf{V}_h = \mathbf{\Sigma}_h^{-1}(\mathbf{\Sigma}_h-\mathbf{S}_h)\mathbf{\Sigma}_h^{-1}\mathbf{\Lambda}\mathbf{\Phi}_h \qquad (h = 1, 2),$$

$$\partial F/\partial\mathbf{\Phi}_1 = 2n_1\mathbf{W}_1, \quad (9.7)$$

and

$$\partial F/\partial\mathbf{\Phi}_2 = 2n_2\mathbf{W}_2-n_2\,\text{diag}\,(\mathbf{W}_2), \quad (9.8)$$

where

$$\mathbf{W}_h = \mathbf{\Lambda}'\mathbf{\Sigma}_h^{-1}(\mathbf{\Sigma}_h-\mathbf{S}_h)\mathbf{\Sigma}_h^{-1}\mathbf{\Lambda}.$$

We also have

$$\partial F/\partial\mathbf{\Psi} = n_1\,\text{diag}\,(\mathbf{U}_1)+n_2\,\text{diag}\,(\mathbf{U}_2), \quad (9.9)$$

where

$$\mathbf{U}_h = \mathbf{\Sigma}_h^{-1}(\mathbf{\Sigma}_h-\mathbf{S}_h)\mathbf{\Sigma}_h^{-1}.$$

In (9.6) elements of $\partial F/\partial\mathbf{\Lambda}$ corresponding in position to fixed parameters in $\mathbf{\Lambda}$ are understood to be zero, and so are the diagonal elements of $\partial F/\partial\mathbf{\Phi}_1$ in (9.7). The reason for the second term on the right-hand side of (9.8) is that the diagonal elements of $\partial F/\partial\mathbf{\Phi}_2$, which correspond in position to free parameters in $\mathbf{\Phi}_2$, are n_2 times the corresponding diagonal elements of the matrix $\mathbf{W}_2$.

In the minimisation procedure the Fletcher and Powell method may again be used. An initial $\mathbf{E}$ matrix may clearly be constructed in the same way as that of Chapter 7. We omit any further description, however, since no opportunity has yet been found for trying out the procedure on any experimental results.

It may be remarked that if the model were generalised to allow the two populations to have different sets of residual variances, represented by the diagonal matrices $\mathbf{\Psi}_1$ and $\mathbf{\Psi}_2$, we should then have, in place of (9.1),

$$\mathbf{\Sigma}_1 = \mathbf{\Lambda}\mathbf{\Phi}_1\mathbf{\Lambda}' + \mathbf{\Psi}_1$$

and

$$\mathbf{\Sigma}_2 = \mathbf{\Lambda}\mathbf{\Phi}_2\mathbf{\Lambda}' + \mathbf{\Psi}_2.$$

The function to be minimised would take the form $F(\mathbf{\Lambda}, \mathbf{\Phi}_1, \mathbf{\Phi}_2, \mathbf{\Psi}_1, \mathbf{\Psi}_2)$. The first-order derivatives of F with respect to $\mathbf{\Lambda}$, $\mathbf{\Phi}_1$ and $\mathbf{\Phi}_2$ would still be as given by (9.6), (9.7) and (9.8), but in place of (9.9) we should have

$$\partial F/\partial \mathbf{\Psi}_h = n_h \operatorname{diag}(\mathbf{U}_h) \qquad (h = 1, 2).$$

9.4 TESTS OF HYPOTHESES

Let H be the hypothesis regarding $\mathbf{\Sigma}_1$ and $\mathbf{\Sigma}_2$ that is represented by equations (9.1). If the alternative to H is simply that $\mathbf{\Sigma}_1$ and $\mathbf{\Sigma}_2$ are two symmetric and positive definite matrices, then the likelihood ratio method of testing H leads to a criterion

$$U = F(\hat{\mathbf{\Lambda}}, \hat{\mathbf{\Phi}}_1, \hat{\mathbf{\Phi}}_2, \hat{\mathbf{\Psi}}),$$

where circumflex accents now denote maximum likelihood estimates, so that U is merely the minimised value of F. For large samples the distribution of U under H is approximately that of χ^2 with $p(p+1)-q$ degrees of freedom, where q is as given by (9.2). In the trivial case where $k = 0$, and where H is that $\mathbf{\Sigma}_1 = \mathbf{\Sigma}_2 = \mathbf{\Psi}$, the χ^2 approximation is improved if U is multiplied by

$$1-(1/12)(2p+3-1/p)(1/n_1+1/n_2)+1/(3pn).$$

When $k > 0$ the correct multiplying factor is unknown, but a reasonable procedure would be to take it as

$$1-(1/12)(2p+3)(1/n_1+1/n_2),$$

ignoring terms of order $1/p$.

We may also wish to test the hypothesis H^* that the factors have the same covariance matrix $\mathbf{\Phi}$ in both populations. If we replace each of $\mathbf{\Sigma}_1$ and $\mathbf{\Sigma}_2$ by

$$\mathbf{\Sigma} = \mathbf{\Lambda}\mathbf{\Phi}\mathbf{\Lambda}' + \mathbf{\Psi},$$

the function F becomes

$$n \log |\mathbf{\Sigma}| + n \operatorname{tr} (\mathbf{S\Sigma}^{-1}) - n_1 \log |\mathbf{S}_1| - n_2 \log |\mathbf{S}_2| - np, \quad (9.10)$$

where $\mathbf{S}$ is as defined in (9.3). Let U^* denote the result of minimising this expression with respect to $\mathbf{\Lambda}, \mathbf{\Phi}$ and $\mathbf{\Psi}$ (the minimisation procedure being equivalent to that of Chapter 7). Then to test H^* against H as alternative we may use the criterion $(U^* - U)$. Under H^* this is distributed for large samples approximately as χ^2 with $\frac{1}{2}k(k+1)$ degrees of freedom. If $\hat{\mathbf{\Sigma}}_1$ and $\hat{\mathbf{\Sigma}}_2$ are the values of $\mathbf{\Sigma}_1$ and $\mathbf{\Sigma}_2$ for which (9.5) is minimised and if $\hat{\mathbf{\Sigma}}$ is the value of $\mathbf{\Sigma}$ for which (9.10) is minimised, the criterion may alternatively be expressed as

$$n \log |\hat{\mathbf{\Sigma}}| - n_1 \log |\hat{\mathbf{\Sigma}}_1| - n_2 \log |\hat{\mathbf{\Sigma}}_2|. \quad (9.11)$$

9.5 ESTIMATING FACTOR SCORES FOR TWO POPULATIONS

Suppose that we wish to estimate factor scores for individuals belonging to either of the two populations. Then it seems best to use the Bartlett method of estimation described in section 8.4, since the matrix on the right-hand side of (8.11) depends only on $\mathbf{\Lambda}$ and $\mathbf{\Psi}$, and not on the factor covariance matrix. Thus the linear functions of the x-variates that estimate the factor scores are the same for both populations. This would not be the case if the regression estimates of section 8.3 were employed, unless $\mathbf{\Phi}_1$ and $\mathbf{\Phi}_2$ happened to be equal. Even with Bartlett's method of estimating factor scores, difficulties would arise if the populations were allowed to have different sets of residual variances, represented by $\mathbf{\Psi}_1$ and $\mathbf{\Psi}_2$, since this would also result in different linear functions being used as estimators in the two populations.

Assuming that $\mathbf{\Psi}$ is invariant, we may test whether the estimated factor scores have the same covariance matrix in both populations. Instead of working with the vector variate $\hat{\mathbf{f}}^*$, as given by (8.11), it is more convenient to use

$$\mathbf{y} = \mathbf{\Gamma}\hat{\mathbf{f}}^* = \mathbf{\Lambda}'\mathbf{\Psi}^{-1}\mathbf{x}.$$

We therefore test the hypothesis H_0 that the vector variate $\mathbf{y}$, of order k, has the same covariance matrix in both populations. The problem of devising such a test is somewhat intractable unless errors of estimation in $\mathbf{\Lambda}$ and $\mathbf{\Psi}$ are ignored. We therefore assume that the elements of $\mathbf{y}$ are known linear functions of the elements of $\mathbf{x}$. This is to some extent justifiable since we are using these linear functions merely for the purpose of comparison.

Given random samples from the two populations, as before, the two sample covariance matrices of $\mathbf{y}$ are respectively $\mathbf{Z}_1$ and $\mathbf{Z}_2$, where

$$\mathbf{Z}_h = \mathbf{\Lambda}'\mathbf{\Psi}^{-1}\mathbf{S}_h\mathbf{\Psi}^{-1}\mathbf{\Lambda}.$$

The hypothesis H_0 may be tested by use of the criterion

$$n \log|\mathbf{Z}| - n_1 \log|\mathbf{Z}_1| - n_2 \log|\mathbf{Z}_2|, \qquad (9.12)$$

where

$$\mathbf{Z} = (1/n)(n_1\mathbf{Z}_1 + n_2\mathbf{Z}_2).$$

Under H_0 this is distributed for large samples approximately as χ^2 with $\frac{1}{2}k(k+1)$ degrees of freedom. The approximation is in theory improved if the criterion is multiplied by the factor

$$1-(1/6)[2k+1-2/(k+1)](1/n_1+1/n_2-1/n),$$

in accordance with a result of Box (1949), though this is strictly justifiable only if $\mathbf{\Lambda}$ and $\mathbf{\Psi}$ are given their true values. The above test may be regarded as an alternative to that of (9.11), from which it differs slightly. In practice both tests would usually lead to the same conclusion.

Other hypotheses might well be of interest. For example, it might be useful to test whether the factor scores had the same set of means in both populations. This could be accomplished by examining the differences of their sample means. However, it hardly seems desirable to discuss this and other problems in any detail since at present little is known as to how well the model that we have proposed would fit actual data.

Appendix I

MATRIX ALGEBRA

A1.1 DEFINITION OF A MATRIX

A $p \times q$ matrix **A** is a rectangular array of numbers, or elements, arranged in p rows and q columns. In full it would be written as

$$\mathbf{A} = \begin{bmatrix} a_{11} & a_{12} & \cdot & \cdot & \cdot & a_{1q} \\ a_{21} & a_{22} & \cdot & \cdot & \cdot & a_{2q} \\ \cdot & \cdot & & & & \cdot \\ \cdot & \cdot & & & & \cdot \\ \cdot & \cdot & & & & \cdot \\ a_{p1} & a_{p2} & \cdot & \cdot & \cdot & a_{pq} \end{bmatrix}.$$

This is usually abbreviated to $\mathbf{A} = [a_{ij}]$, where $i = 1, \ldots, p$ and $j = 1, \ldots, q$. Throughout this book and in this appendix we confine our attention to matrices whose elements are real numbers. Capital boldface letters are used to denote matrices. Their elements are denoted by the corresponding italic lower-case letters with appropriate suffices.

A1.2 OPERATIONS ON MATRICES

The sum of two matrices **A** and **B**, with the same numbers of rows and columns, is defined by

$$\mathbf{A}+\mathbf{B} \equiv [a_{ij}]+[b_{ij}] = [a_{ij}+b_{ij}].$$

The product of a matrix **A** and a real number, or scalar, λ is defined by

$$\lambda\mathbf{A} = \mathbf{A}\lambda = [\lambda a_{ij}].$$

If $\lambda = -1$, the product is written as $-\mathbf{A}$. A null or zero matrix is one whose elements are all zeros. If **A** is null, we write $\mathbf{A} = \mathbf{0}$. It may be verified that these operations satisfy the laws

$$\mathbf{A}+\mathbf{B} = \mathbf{B}+\mathbf{A},$$
$$(\mathbf{A}+\mathbf{B})+\mathbf{C} = \mathbf{A}+(\mathbf{B}+\mathbf{C}),$$
$$\mathbf{A}-\mathbf{A} = \mathbf{0},$$
$$(\lambda+\mu)\mathbf{A} = \lambda\mathbf{A}+\mu\mathbf{A},$$
$$\lambda(\mathbf{A}+\mathbf{B}) = \lambda\mathbf{A}+\lambda\mathbf{B}.$$

If **A** has as many columns as **B** has rows, it is possible to define the product **AB**. Suppose that $\mathbf{A} = [a_{ij}]$ is of order $p \times q$ and that $\mathbf{B} = [b_{ij}]$ is of order $q \times r$. Then $\mathbf{AB} = \mathbf{C}$ is a matrix of order $p \times r$. The element c_{ik} in the ith row and kth column of **C** is given by

$$c_{ik} = \sum_{j=1}^{q} a_{ij}b_{jk}.$$

It will be noted that this is the sum of products of corresponding elements in the ith row of **A** and the kth column of **B**.

The product **BA** may be meaningless even if **AB** exists. The two products can co-exist only if, for **A** of order $p \times q$, **B** is of order $q \times p$. In this case **AB** is a square matrix of order p and **BA** is a square matrix of order q. Even if $p = q$, and both **A** and **B** are square, it can easily be verified that **AB** is not in general equal to **BA**. We must therefore always distinguish between pre-multiplication and post-multiplication. In the product **AB**, for example, **A** is post-multiplied by **B**, while **B** is pre-multiplied by **A**.

If **A**, **B** and **C** are matrices of order $p \times q$, $q \times r$ and $r \times s$ respectively, it is possible to form the product **AB** and to post-multiply it by **C**. The result may be written as (**AB**)**C**. We can also find the product **BC** and pre-multiply it by **A** to obtain a matrix **A**(**BC**). From the definition of matrix multiplication it is clear that

$$(\mathbf{AB})\mathbf{C} = \mathbf{A}(\mathbf{BC}).$$

Hence each of these two matrices may be denoted without ambiguity by **ABC**. Similar results hold for products of more than three matrices. In finding such products it is important that the correct order of the matrices should be preserved. For a square matrix **A** it is convenient to write $\mathbf{A}^2 = \mathbf{AA}$, $\mathbf{A}^3 = \mathbf{AAA}$, etc.

It may be noted that matrix multiplication also satisfies the laws

$$\mathbf{A}(\mathbf{B}+\mathbf{C}) = \mathbf{AB}+\mathbf{AC},$$
$$(\mathbf{A}+\mathbf{B})\mathbf{C} = \mathbf{AC}+\mathbf{BC}.$$

Thus matrices obey all the ordinary laws of elementary algebra except the commutative law of multiplication.

The transpose of the $p \times q$ matrix $\mathbf{A} = [a_{ij}]$ is defined to be the $q \times p$ matrix $\mathbf{A}' = [a'_{ij}]$ for which $a'_{ij} = a_{ji}$. Thus the element in the *i*th row and *j*th column of $\mathbf{A}'$ is the same as the element in the *j*th row and *i*th column of $\mathbf{A}$. (Some authors use a T as affix instead of a prime.) The operation of transposition has the properties

$$(\mathbf{A}')' = \mathbf{A},$$
$$(\mathbf{A}+\mathbf{B})' = \mathbf{A}'+\mathbf{B}',$$
$$(\mathbf{AB})' = \mathbf{B}'\mathbf{A}'.$$

For the transpose of the product of more than two matrices the rule is that the transposed matrices are multiplied in reverse order. Thus, for example,

$$(\mathbf{ABC})' = \mathbf{C}'\mathbf{B}'\mathbf{A}'.$$

A1.3 ROW AND COLUMN VECTORS

Matrices consisting of only one row or column are termed vectors. We use boldface lower-case letters to denote these. For clarity such letters bear a prime if a row vector is intended, but no prime if a column vector is intended. We write, for example,

$$\mathbf{x}' = [x_1 x_2 \ldots x_p].$$

The corresponding column vector may be written in full as

$$\mathbf{x} = \{x_1 x_2 \ldots x_p\},$$

where the curled brackets here indicate that a vertical alignment of the elements is intended. This convention is not, however, in general use. The vector $\mathbf{x}$, having p elements, is said to be of order p.

Let $\mathbf{y}$ be another vector of order p having elements $y_1, \ldots, y_p$. If we pre-multiply it by $\mathbf{x}'$ we obtain a 1×1 matrix containing only one element. The product is thus a scalar and we have

$$\mathbf{x}'\mathbf{y} = \mathbf{y}'\mathbf{x} = x_1 y_1 + x_2 y_2 + \ldots + x_p y_p.$$

A quantity of this kind is often termed either the scalar product or the inner product of the two vectors. Note, however, that $\mathbf{xy}'$ and $\mathbf{yx}'$ are both $p \times p$ matrices.

The square root of the non-negative scalar

$$\mathbf{x}'\mathbf{x} = x_1^2 + x_2^2 + \ldots + x_p^2$$

is often termed the length of $\mathbf{x}$. Provided that $\mathbf{x}$ is non-null it can be scaled to unit length by multiplication by the scalar $1/\sqrt{(\mathbf{x}'\mathbf{x})}$. I $\mathbf{x}'\mathbf{x} = 1$, the vector $\mathbf{x}$ is said to be standardised, or normalised.

A1.4 SOME SPECIAL MATRICES

From now on we shall be concerned, unless we state otherwise, with square matrices of the same order p, which can be added and multiplied at will, and with vectors of order p.

The matrix $\mathbf{A}$ is said to be symmetric if $\mathbf{A} = \mathbf{A}'$, or $a_{ij} = a_{ji}$ for all i and j. It may be noted that a symmetric matrix of order p has, in general, $\frac{1}{2}p(p+1)$ distinct elements.

A diagonal matrix is one whose non-diagonal elements are all zero. Thus if $\mathbf{A}$ is diagonal, $a_{ij} = 0$ unless $i = j$.

A diagonal matrix of special interest is the unit matrix $\mathbf{I}$. This has unities as its diagonal elements. If it is important to indicate the order of the matrix we write it as $\mathbf{I}_p$, but otherwise the suffix is omitted. The unit matrix satisfies

$$\mathbf{IA} = \mathbf{AI} = \mathbf{A}.$$

For some purposes we require triangular matrices. A lower triangular matrix $\mathbf{T} = [t_{ij}]$ is such that all elements above the diagonal are zero. Thus $t_{ij} = 0$ if $i < j$. The transposed matrix $\mathbf{T}'$ is said to be upper triangular, since all elements of $\mathbf{T}'$ below the diagonal are zero. It may easily be verified that the product of two lower triangular matrices is also lower triangular. A similar property holds for upper triangular matrices.

The matrix $\mathbf{A}$ is said to be of tri-diagonal form if $a_{ij} = 0$ unless $|i-j| < 2$. In this case only diagonal elements of $\mathbf{A}$ or elements adjacent to them can be non-zero. Tri-diagonal matrices occur in the evaluation by digital computers of the latent roots and vectors of a symmetric matrix (*see* section A1.12).

A1.5 PARTITIONING OF MATRICES

It is sometimes convenient to represent a matrix in a partitioned form by the juxtaposition of two or more submatrices. An example of a partitioned matrix $\mathbf{A}$, not necessarily square, is

$$\mathbf{A} = \begin{bmatrix} \mathbf{P} & \mathbf{Q} \\ \mathbf{R} & \mathbf{S} \end{bmatrix},$$

where $\mathbf{P}$ and $\mathbf{Q}$ are matrices having the same number of rows, $\mathbf{P}$ and $\mathbf{R}$ have the same number of columns, and so on. If there is any danger of confusion, the submatrices may be separated by dotted lines. From the definition of a transpose it follows that

$$\mathbf{A}' = \begin{bmatrix} \mathbf{P}' & \mathbf{R}' \\ \mathbf{Q}' & \mathbf{S}' \end{bmatrix}.$$

Suppose that in the partitioned matrix

$$\mathbf{B} = \begin{bmatrix} \mathbf{W} & \mathbf{X} \\ \mathbf{Y} & \mathbf{Z} \end{bmatrix},$$

not necessarily square, the submatrices **W** and **X** have as many rows as **P** and **R** have columns and that **Y** and **Z** have as many rows as **Q** and **S** have columns. Then the product **AB** exists and is given by

$$\mathbf{AB} = \begin{bmatrix} \mathbf{P} & \mathbf{Q} \\ \mathbf{R} & \mathbf{S} \end{bmatrix} \begin{bmatrix} \mathbf{W} & \mathbf{X} \\ \mathbf{Y} & \mathbf{Z} \end{bmatrix}$$
$$= \begin{bmatrix} \mathbf{PW}+\mathbf{QY} & \mathbf{PX}+\mathbf{QZ} \\ \mathbf{RW}+\mathbf{SY} & \mathbf{RX}+\mathbf{SZ} \end{bmatrix}.$$

Thus the usual rule of matrix multiplication is applied with submatrices of **A** and **B** treated as if they were elements.

A1.6 DETERMINANTS, MINORS AND COFACTORS

For any square matrix **A** we can define the determinant, written $|\mathbf{A}|$ or $|a_{ij}|$, by

$$|\mathbf{A}| = \sum \pm a_{1\alpha} a_{2\beta} \ldots a_{p\nu},$$

where the summation, of $p!$ terms, is over all permutations α, β, ..., ν of the integers $1, 2, \ldots, p$. The sign $+$ or $-$ is prefixed to each term according as the permutation is even or odd. To explain the terms even and odd in this connection we note that the integers $1, 2, \ldots, p$, in natural order, may be changed into a different order $\alpha, \beta, \ldots, \nu$ by a number of successive transpositions, where a transposition means the interchange in position of two integers. A particular permutation $\alpha, \beta, \ldots, \nu$ may be obtained by many different sequences of transpositions, but it can be shown that the number of transpositions performed is either always even or always odd. The sign of the corresponding term in the summation is $+$ if the number is even and $-$ if the number is odd.

If **A** is either a diagonal or a triangular matrix, the determinant is given by

$$|\mathbf{A}| = a_{11} a_{22} \ldots a_{pp}.$$

In general, however, the above definition is unsuitable for the numerical evaluation of determinants if $p > 3$, and other methods are used (*see* section A1.15).

It can be shown that $|\mathbf{A}'| = |\mathbf{A}|$ and that

$$|\mathbf{AB}| = |\mathbf{BA}| = |\mathbf{A}| \times |\mathbf{B}|.$$

Similar results hold for the determinant of the product of any number of square matrices of the same order.

From **A** we can form submatrices by deleting a number of rows and columns. A minor is the determinant of a square submatrix of **A**. The minor of a_{ij} in $|\mathbf{A}|$ is the determinant of the submatrix of **A** found by deleting the ith row and the jth column. The cofactor of a_{ij} in $|\mathbf{A}|$, denoted by A_{ij}, is $(-1)^{i+j}$ times the minor of a_{ij}. It can be shown that

$$\sum_k a_{ik}A_{ik} = \sum_k a_{ki}A_{ki} = |\mathbf{A}| \tag{A1.1}$$

and that, if $i \neq j$,

$$\sum_k a_{ik}A_{jk} = \sum_k a_{ki}A_{kj} = 0. \tag{A1.2}$$

If $p = 3$, equation (A1.1) is useful for the numerical evaluation of $|\mathbf{A}|$. For $p > 3$ the calculation of the cofactors requires too much laborious arithmetic.

A1.7 INVERSE OF A MATRIX

If $|\mathbf{A}| = 0$ the matrix **A** is said to be singular. Let us suppose, however, that $|\mathbf{A}| \neq 0$ and that **A** is non-singular. We may then define a matrix $\mathbf{A}^{-1} = [a^{ij}]$ by

$$a^{ij} = A_{ji}/|\mathbf{A}| \qquad (i, j = 1, \ldots, p). \tag{A1.3}$$

Equations (A1.1) and (A1.2) may be rewritten as

$$\sum_k a_{ik}a^{ki} = \sum_k a^{ik}a_{ki} = 1$$

and, for $i \neq j$,

$$\sum_k a_{ik}a^{kj} = \sum_k a^{ik}a_{kj} = 0.$$

These equations when expressed in terms of matrices become simply

$$\mathbf{AA}^{-1} = \mathbf{A}^{-1}\mathbf{A} = \mathbf{I}.$$

The matrix $\mathbf{A}^{-1}$ is termed the inverse of **A**. The above definition shows that it exists and that it is unique provided that **A** is non-singular. For $p > 3$ equation (A1.3) is unsuitable for the numerical evaluation of $\mathbf{A}^{-1}$. Other more convenient methods are discussed in section A1.15.

Since

$$|\mathbf{A}||\mathbf{A}^{-1}| = |\mathbf{A}\mathbf{A}^{-1}| = |\mathbf{I}| = 1,$$

we have

$$|\mathbf{A}^{-1}| = 1/|\mathbf{A}|.$$

It is easy to show that $(\mathbf{A}')^{-1} = (\mathbf{A}^{-1})'$. If **A** and **B** are both non-singular, then $(\mathbf{AB})^{-1}$ exists and is equal to $\mathbf{B}^{-1}\mathbf{A}^{-1}$. This is so because

$$(\mathbf{AB})(\mathbf{B}^{-1}\mathbf{A}^{-1}) = \mathbf{A}(\mathbf{B}\mathbf{B}^{-1})\mathbf{A}^{-1} = \mathbf{AIA}^{-1} = \mathbf{AA}^{-1} = \mathbf{I}.$$

Similarly, if **A**, **B** and **C** are all non-singular,

$$(\mathbf{ABC})^{-1} = \mathbf{C}^{-1}\mathbf{B}^{-1}\mathbf{A}^{-1}.$$

The rule for finding the inverse of the product of any number of non-singular matrices is that the inverse matrices must be multiplied together in reverse order. If **D** is a diagonal matrix with elements $d_1, \ldots, d_p$, then $\mathbf{D}^{-1}$ is diagonal with elements $1/d_1, \ldots, 1/d_p$.

A1.8 LINEAR EQUATIONS AND TRANSFORMATIONS

For simultaneous linear equations in several variables the notation of matrix algebra is particularly suitable. The set of p equations

$$a_{i1}x_1 + a_{i2}x_2 + \ldots + a_{ip}x_p = c_i \qquad (i = 1, \ldots, p)$$

in the p unknown variables $x_1, \ldots, x_p$ can be written concisely as

$$\mathbf{Ax} = \mathbf{c}.$$

Suppose that **A** is non-singular and that $\mathbf{A}^{-1}$ has been found. Pre-multiplication of the above equation by $\mathbf{A}^{-1}$ then gives the solution in the form

$$\mathbf{x} = \mathbf{A}^{-1}\mathbf{c}$$

and shows that it is unique.

It is often necessary to transform linearly one set of variables or co-ordinates $x_1, \ldots, x_p$ into another set $y_1, \ldots, y_p$. The transformation giving the y's in terms of the x's and the inverse transformation giving the x's in terms of the y's may be represented by the respective equations

$$\mathbf{y} = \mathbf{Ax}, \qquad \mathbf{x} = \mathbf{A}^{-1}\mathbf{y},$$

where **A** is non-singular.

A1.9 ORTHOGONAL MATRICES AND TRANSFORMATIONS

In many cases both the x's and the y's represent rectangular, or orthogonal, co-ordinates. Both the transformation and the matrix defining it are then termed orthogonal. An orthogonal matrix **A** satisfies

$$\mathbf{AA}' = \mathbf{A}'\mathbf{A} = \mathbf{I},$$

and thus

$$\mathbf{A}' = \mathbf{A}^{-1}, \qquad \mathbf{A} = (\mathbf{A}')^{-1}.$$

In view of this, an orthogonal transformation is represented by

$$\mathbf{y} = \mathbf{Ax}, \qquad \mathbf{x} = \mathbf{A}'\mathbf{y}.$$

It may be noted that

$$\mathbf{y}'\mathbf{y} = (\mathbf{x}'\mathbf{A}')(\mathbf{Ax}) = \mathbf{x}'(\mathbf{A}'\mathbf{A})\mathbf{x} = \mathbf{x}'\mathbf{Ix} = \mathbf{x}'\mathbf{x}.$$

Considered geometrically this expresses the fact that the square of the distance of a point from the origin remains invariant under a change of co-ordinate axes.

For an orthogonal matrix **A** we have

$$|\mathbf{A}|^2 = |\mathbf{A}|\,|\mathbf{A}'| = |\mathbf{AA}'| = |\mathbf{I}| = 1,$$

and thus $|\mathbf{A}| = \pm 1$. The sign can always be made + by changing, if necessary, the signs of all elements in any row or column of **A**. Such sign changes clearly do not affect the orthogonality of **A**.

Any orthogonal matrix of order 2 whose determinant is +1 may be expressed in the form

$$\begin{bmatrix} \cos\theta & \sin\theta \\ -\sin\theta & \cos\theta \end{bmatrix}.$$

The transformation of co-ordinates from (x_1, x_2) to (y_1, y_2) represents a rotation through an angle θ.

A1.10 LINEAR INDEPENDENCE AND RANK

A set of vectors $\mathbf{z}_1, \ldots, \mathbf{z}_k$ is said to be linearly independent if there exists no set of scalars $c_1, \ldots, c_k$, not all zero, such that

$$c_1\mathbf{z}_1 + \ldots + c_k\mathbf{z}_k = \mathbf{0}.$$

A matrix **A**, rectangular or square, is said to be of rank r if the maximum number of linearly independent rows (or columns) is r. Alternatively and equivalently, the rank of **A** is defined to be r if every minor of order $r+1$ formed from **A** is zero and at least one minor or order r is not zero. If the rank of a square matrix **A** is less than its order, **A** is singular, and conversely.

A1.11 LATENT ROOTS AND VECTORS

The latent roots (sometimes called characteristic roots or eigenvalues) of a matrix **A** are the values of λ satisfying

$$|\mathbf{A}-\lambda\mathbf{I}| = 0, \qquad \text{(A1.4)}$$

which is an equation for λ of degree p. Let us suppose that the latent roots are distinct and that they are arranged in some convenient order. Corresponding to the rth root λ_r $(r = 1, \ldots, p)$ there is a column vector $\mathbf{u}_r$ satisfying

$$\mathbf{A}\mathbf{u}_r = \lambda_r\mathbf{u}_r. \qquad \text{(A1.5)}$$

This is known as the rth latent column vector (or eigenvector) of **A**. Similarly there is a latent row vector $\mathbf{v}_r'$ satisfying

$$\mathbf{v}_r'\mathbf{A} = \lambda_r\mathbf{v}_r'.$$

The vectors are unique except that each may be multiplied by an arbitrary scalar, either positive or negative.

Suppose now that **A** is symmetric. Then all its latent roots are real and, since we have assumed them to be distinct, we may suppose that they are arranged in decreasing order of magnitude, the largest root being λ_1. Transposition of (A1.5) shows that the latent row vector $\mathbf{v}_r'$ is simply the transpose of the column vector $\mathbf{u}_r$.

For convenience we may standardise each vector, so that $\mathbf{u}_r'\mathbf{u}_r = 1$ for each value of r. The latent vectors are then uniquely determined apart from possible multiplications by -1. It can be shown that $\mathbf{u}_r'\mathbf{u}_s = 0$ if $\mathbf{r} \neq \mathbf{s}$. In view of these conditions the $p \times p$ matrix **U** whose columns are $\mathbf{u}_1, \ldots, \mathbf{u}_p$, is orthogonal.

Let $\mathbf{\Lambda}$ be the diagonal matrix of order p whose diagonal elements are $\lambda_1, \ldots, \lambda_p$. Then the two sides of (A1.5) are the rth columns of the respective matrices **AU** and $\mathbf{U\Lambda}$. Hence we have the equation

$$\mathbf{AU} = \mathbf{U\Lambda}. \qquad \text{(A1.6)}$$

Pre-multiplication of this by $\mathbf{U}'$ gives

$$\mathbf{\Lambda} = \mathbf{U}'\mathbf{AU}, \qquad \text{(A1.7)}$$

while post-multiplication by $\mathbf{U}'$ gives

$$\mathbf{A} = \mathbf{U\Lambda U}'. \qquad \text{(A1.8)}$$

From this equation we have

$$|\mathbf{A}| = |\mathbf{U}|\,|\mathbf{\Lambda}|\,|\mathbf{U}'| = |\mathbf{\Lambda}|,$$

and hence

$$|\mathbf{A}| = \lambda_1\lambda_2\ldots\lambda_p.$$

The rank of **A** is equal to the number of non-zero latent roots. If none of them is zero, **A** is non-singular, and conversely. In that case we can invert (A1.8) to give

$$\mathbf{A}^{-1} = \mathbf{U}\mathbf{\Lambda}^{-1}\mathbf{U}',$$

which may be used to find the inverse of a symmetric matrix whose latent roots and vectors are known. The latent vectors of $\mathbf{A}^{-1}$ are the same as those of **A**; the roots are $1/\lambda_1, \ldots, 1/\lambda_p$.

Suppose that **A** is a $p \times q$ matrix and that **B** is a $q \times p$ matrix. It can be shown that if $p > q$, the latent roots of **AB** are those of **BA** together with a zero root of multiplicity $p-q$, and similarly for the case where $q > p$. If $p = q$, the two sets of roots are identical. It follows that the latent roots of $\mathbf{I}_p + \mathbf{AB}$ are the same as those of $\mathbf{I}_q + \mathbf{BA}$ except possibly for a number of unities. Hence we have

$$|\mathbf{I}_p + \mathbf{AB}| = |\mathbf{I}_q + \mathbf{BA}|, \tag{A1.9}$$

a result that is made use of in Chapter 7.

A1.12 NUMERICAL EVALUATION OF LATENT ROOTS AND VECTORS

It is often necessary to evaluate numerically the latent roots and vectors of a symmetric matrix **A**. Before the advent of electronic computers an iterative method was often used for doing this. Starting with a trial vector **u**, the sequence of vectors

$$\mathbf{Au}, \mathbf{A}^2\mathbf{u}, \mathbf{A}^3\mathbf{u}, \ldots$$

is found and scaled in some convenient way. Provided that the choice of **u** has not been unlucky, the sequence eventually converges to the first latent vector $\mathbf{u}_1$. The largest latent root λ_1 is then given by $(\mathbf{u}_1'\mathbf{A}\mathbf{u}_1)/(\mathbf{u}_1'\mathbf{u}_1)$.

To find the second latent root and vector the procedure is now repeated with $\mathbf{A} - \lambda_1\mathbf{u}_1\mathbf{u}_1'$ in place of **A**. Subsequent roots and vectors are found in a similar manner.

Convergence with this procedure is usually rather slow and, though there are methods for accelerating it, a considerable amount of calculation is required for matrices of moderately large order.

In computer programs an entirely different method of evaluating the latent roots and vectors is used. Here it is possible to give only a very brief indication of the way in which this method works. A full account has been given by Wilkinson (1965).

Let us first note that if **w** is a standardised vector, $\mathbf{P} = \mathbf{I} - 2\mathbf{ww}'$ is a symmetric orthogonal matrix satisfying $\mathbf{P}^2 = \mathbf{PP}' = \mathbf{I}$. Consider the transformation in which **A** is replaced by **PAP**. The latter matrix is symmetric and has the same latent roots as **A**. Its latent column vectors are those of **A** pre-multiplied by **P**.

The first step is to reduce **A** to a new symmetric matrix of tri-diagonal form. This is best achieved by the method of Householder, in which $p-2$ successive transformations of the above type are carried out. By suitable choices of the **P**-matrices involved the final matrix **B** is made to be symmetric and tri-diagonal. The problem is thus reduced to that of finding the latent roots and vectors of a matrix of this type. From the latent vectors of **B**, when found, we can transform back to those of **A**.

With a computer the latent roots of a symmetric tri-diagonal matrix may easily be obtained. Let μ be any trial value and define $g_p = g_p(\mu)$ as the determinant $|\mathbf{B}-\mu\mathbf{I}|$. For $r = 1, \ldots, p-1$, let $g_r = g_r(\mu)$ be the minor formed from this determinant by retaining only its first r rows and columns and deleting the rest. Taking $g_0 = 1$, we then have a sequence $g_0, g_1, \ldots, g_p$. Omit any zeros that occur and consider pairs of consecutive non-zero members of the sequence. Let $n(\mu)$ be the number of pairs in which the members are of opposite sign. It can be shown that $n(\mu)$ is the number of latent roots that are less than μ. By making use of this result all p roots may be found with a high degree of accuracy. The method of computation, though iterative, is extremely rapid.

Let λ denote a value near, but not very close to, a latent root. Then the corresponding vector may be found by solving successively the equations

$$(\mathbf{B}-\lambda\mathbf{I})\mathbf{x} = \mathbf{v},$$
$$(\mathbf{B}-\lambda\mathbf{I})\mathbf{y} = \mathbf{x},$$

where **v** is an arbitrary vector. If these equations are solved in the most efficient manner, the vector **y** is usually a very good approximation to the required latent vector. Difficulties arise only if two or more roots coincide or are close together. The value of λ must not be an exact root as $\mathbf{B}-\lambda\mathbf{I}$ would then be a singular matrix and the equations for **x** and **y** could not be solved.

A1.13 QUADRATIC FORMS

The expression

$$\mathbf{x}'\mathbf{A}\mathbf{x} = \sum_{i,j} a_{ij}x_i x_j,$$

where $\mathbf{A} = [a_{ij}]$ is symmetric, is known as a quadratic form. Both it and the matrix **A** are termed positive definite if $\mathbf{x}'\mathbf{A}\mathbf{x} > 0$ for all non-null vectors **x**. If $>$ is replaced by $\geqslant$, they are termed either non-negative definite or positive semi-definite.

With **U** and **Λ** as in section 1.11, consider the orthogonal transformation defined by

$$\mathbf{y} = \mathbf{U}'\mathbf{x}, \qquad \mathbf{x} = \mathbf{U}\mathbf{y}.$$

Under this transformation we have, using (A1.7),

$$\mathbf{x}'\mathbf{A}\mathbf{x} = \mathbf{y}'\mathbf{U}'\mathbf{A}\mathbf{U}\mathbf{y} = \mathbf{y}'\mathbf{\Lambda}\mathbf{y},$$

or, in terms of the elements of **y**,

$$\mathbf{x}'\mathbf{A}\mathbf{x} = \sum_{r=1}^{p} \lambda_r y_r^2.$$

From this expression it is clear that a necessary and sufficient condition for **A** to be positive definite is that all its latent roots should be positive. For **A** to be non-negative definite the condition is that all roots should be either positive or zero. If **A** is positive definite, then $\mathbf{A}^{-1}$ exists and is also positive definite.

A1.14 TRACE OF A MATRIX

The trace of a matrix **A**, written tr **A**, is the sum of its diagonal elements. It may easily be verified that

$$\begin{aligned} \operatorname{tr} \mathbf{A}' &= \operatorname{tr} \mathbf{A}, \\ \operatorname{tr} (\mathbf{A}+\mathbf{B}) &= \operatorname{tr} \mathbf{A} + \operatorname{tr} \mathbf{B}, \\ \operatorname{tr} (\mathbf{AB}) &= \operatorname{tr} (\mathbf{BA}). \end{aligned}$$

The last equation holds even when **A** and **B** are not square provided that both products exist. Note that $\operatorname{tr} (\mathbf{A}'\mathbf{A}) = \operatorname{tr} (\mathbf{A}\mathbf{A}')$ is the sum of squares of the elements of **A**.

From equation (A1.8) we have

$$\operatorname{tr} (\mathbf{A}) = \operatorname{tr} (\mathbf{U\Lambda U}') = \operatorname{tr} (\mathbf{\Lambda U}'\mathbf{U}) = \operatorname{tr} (\mathbf{\Lambda}).$$

Thus tr **A** is the sum of the latent roots of **A**, a result that can also be obtained by expansion of equation (A1.4).

A1.15 NUMERICAL INVERSION OF MATRICES

The numerical inversion of matrices of large order is nowadays carried out by means of electronic computers. If the order is moderately small, however, a desk calculating machine may conveniently be used. Various methods of inverting matrices are available. Those that we give below are suitable both for desk machines and for computers.

First suppose that **A** is symmetric and positive definite. We begin by expressing it in the form **TT**′ where $\mathbf{T} = [t_{ij}]$ is a lower triangular matrix with positive diagonal elements. This is accomplished by equating corresponding elements of **TT**′ and **A**. For the diagonal elements of **A** we have

$$a_{ii} = \sum_{k=1}^{i} t_{ik}^2 \qquad (i = 1, \ldots, p),$$

and for the non-diagonal elements we have

$$a_{ij} = \sum_{k=1}^{j} t_{ik} t_{jk} \qquad (i > j).$$

On transforming these equations we find that

$$t_{11} = \sqrt{a_{11}},$$

$$t_{ii} = \sqrt{(a_{ii} - \sum_{k=1}^{i-1} t_{ik}^2)} \qquad (i > 1),$$

$$t_{i1} = (1/t_{11}) a_{i1} \qquad (i > 1),$$

$$t_{ij} = (1/t_{jj})\,(a_{ij} - \sum_{k=1}^{j-1} t_{ik} t_{jk}) \qquad (i > j > 1).$$

These equations enable us to calculate successively

$$t_{11}, t_{21}, \ldots, t_{p1}, t_{22}, t_{32}, \ldots, t_{p2}, \ldots, t_{pp}.$$

Note that when **T** has been found the determinant of **A** may readily be calculated since

$$|\mathbf{A}| = |\mathbf{T}|^2 = t_{11}^2 t_{22}^2 \ldots t_{pp}^2.$$

The matrix $\mathbf{T}^{-1} = [t^{ij}]$, which is also lower triangular, may be found by equating corresponding elements of $\mathbf{TT}^{-1}$ and **I**. We have

$$t^{ii} = 1/t_{ii} \qquad (i = 1, \ldots, p)$$

and

$$\sum_{k=j}^{i} t_{ik} t^{kj} = 0 \qquad (i > j).$$

The last equation may be put in the form

$$t^{ij} = -(1/t_{ii}) \sum_{k=j}^{i-1} t_{ik} t^{kj} \qquad (i > j),$$

which enables us to calculate successively

$$t^{11}, t^{21}, \ldots, t^{p1}, t^{22}, t^{32}, \ldots, t^{p2}, \ldots, t^{pp}.$$

Finally $\mathbf{A}^{-1}$ is given as $(\mathbf{T}^{-1})'\mathbf{T}^{-1}$.

Suppose now that **A** is any square non-singular matrix. We express it this time in the form $\mathbf{T}_1\mathbf{T}_2'$, where $\mathbf{T}_1$ and $\mathbf{T}_2$ are both lower triangular, the former having unit diagonal elements. The matrices $\mathbf{T}_1$ and $\mathbf{T}_2$ are found by equating corresponding elements of $\mathbf{T}_1\mathbf{T}'_2$ and **A**. Then $\mathbf{A}^{-1}$ is given as $(\mathbf{T}_2^{-1})'\mathbf{T}_1^{-1}$, and $|\mathbf{A}| = |\mathbf{T}_2|$. The diagonal elements of $\mathbf{T}_2$ may be either positive or negative.

The same method may conveniently be used when **A** is symmetric but not necessarily positive definite. Let **D** be the diagonal matrix whose diagonal elements are the same as those of $\mathbf{T}_2$. Then, in this case, $\mathbf{T}_1 = \mathbf{T}_2\mathbf{D}^{-1}$.

A1.16 VECTOR AND MATRIX DIFFERENTIATION

In matrix theory various kinds of derivatives may occur. In discussing them we shall use **X** to denote a variable matrix and **x** to denote a variable vector of order p. On the other hand, **A** will be a constant matrix and **a** will be a constant vector of order p.

Suppose that **x** has elements $x_1, \ldots, x_p$ that are functionally dependent on a scalar variable v. Then $\partial\mathbf{x}/\partial v$ is defined to be the column vector of order p whose ith element is $\partial x_i/\partial v$. Similarly, if the elements of the $p\times q$ matrix **X** are functionally dependent on v, we define $\partial\mathbf{X}/\partial v$ as the $p\times q$ matrix whose element in the ith row and jth column is $\partial x_{ij}/\partial v$.

Assume, first, that there are no functional relationships between the elements of **X** and consider $\partial\mathbf{X}/\partial x_{ij}$. This matrix has a unit element in the ith row and jth column and zeros elsewhere. Now suppose, on the other hand, that **X** is a square and symmetric matrix and that x_{ji} is considered to be identical with x_{ij}. Then if i and j are unequal, $\partial\mathbf{X}/\partial x_{ij}$ has two unit elements, there being an additional one in the jth row and ith column.

The usual rule for differentiating a product applies to matrices provided that the correct order of multiplication is preserved. Thus, if **X** and **Y** are any two variable matrices for which the product **XY** exists, we have

$$\partial(\mathbf{XY})/\partial v = \mathbf{X}(\partial\mathbf{Y}/\partial v)+(\partial\mathbf{X}/\partial v)\mathbf{Y}.$$

Suppose that **X** is square and non-singular. On differentiating the equation $\mathbf{XX}^{-1} = \mathbf{I}$ partially with respect to v we have

$$\mathbf{X}(\partial\mathbf{X}^{-1}/\partial v)+(\partial\mathbf{X}/\partial v)\mathbf{X}^{-1} = \mathbf{0}.$$

Pre-multiplication of this by $\mathbf{X}^{-1}$ then gives

$$\partial\mathbf{X}^{-1}/\partial v = -\mathbf{X}^{-1}(\partial\mathbf{X}/\partial v)\mathbf{X}^{-1}. \qquad \text{(A1.10)}$$

Now let v be a scalar function of the elements of the $p \times q$ matrix $\mathbf{X} = [x_{ij}]$. Then we define $\partial v/\partial \mathbf{X}$ to be the $p \times q$ matrix whose element in the ith row and jth column is $\partial v/\partial x_{ij}$. This definition includes as a special case the vector derivative $\partial v/\partial \mathbf{x}$.

Since

$$\partial(\mathbf{a}'\mathbf{x})/\partial x_i = (\partial/\partial x_i) \sum_j (a_j x_j) = a_i,$$

we have

$$\partial(\mathbf{a}'\mathbf{x})/\partial \mathbf{x} = \partial(\mathbf{x}'\mathbf{a})/\partial \mathbf{x} = \mathbf{a}. \tag{A1.11}$$

Let $\mathbf{A} = [a_{ij}]$ be a symmetric matrix of order p. Then

$$(\partial/\partial x_i)(\mathbf{x}'\mathbf{A}\mathbf{x}) = (\partial/\partial x_i) \sum_{j,k} (a_{jk} x_j x_k)$$
$$= 2\sum_j (a_{ij} x_j).$$

This is the ith element of the vector $2\mathbf{A}\mathbf{x}$ and hence

$$(\partial/\partial \mathbf{x})(\mathbf{x}'\mathbf{A}\mathbf{x}) = 2\mathbf{A}\mathbf{x}. \tag{A1.12}$$

If $\mathbf{X}$ is a $p \times q$ matrix and $\mathbf{A}$ is a $q \times p$ matrix, then

$$(\partial/\partial x_{ij}) \operatorname{tr}(\mathbf{A}\mathbf{X}) = (\partial/\partial x_{ij}) \sum_{h,k} (a_{hk} x_{kh}) = a_{ji},$$

and hence

$$(\partial/\partial \mathbf{X}) \operatorname{tr}(\mathbf{A}\mathbf{X}) = \mathbf{A}'.$$

Henceforth let us suppose that both $\mathbf{A}$ and $\mathbf{X}$ are square and of order p. Let X_{ij} be the cofactor of x_{ij} in $|\mathbf{X}|$. Then since

$$|\mathbf{X}| = \sum_k x_{ik} X_{ik},$$

we have

$$\partial |\mathbf{X}|/\partial x_{ij} = X_{ij}$$

If the inverse matrix $\mathbf{X}^{-1} = [x^{ij}]$ exists, we have

$$(\partial/\partial x_{ij}) \log_e |\mathbf{X}| = |\mathbf{X}|^{-1} \partial |\mathbf{X}|/\partial x_{ij}$$
$$= X_{ij}/|\mathbf{X}|$$
$$= x^{ji},$$

and hence

$$(\partial/\partial \mathbf{X}) \log |\mathbf{X}| = (\mathbf{X}^{-1})'.$$

We have also

$$(\partial/\partial x_{ij}) \operatorname{tr}(\mathbf{A}\mathbf{X}^{-1}) = \operatorname{tr}[\mathbf{A}(\partial \mathbf{X}^{-1}/\partial x_{ij})]$$
$$= \operatorname{tr}[-\mathbf{A}\mathbf{X}^{-1}(\partial \mathbf{X}/\partial x_{ij})\mathbf{X}^{-1}]$$
$$= -\operatorname{tr}[\mathbf{X}^{-1}\mathbf{A}\mathbf{X}^{-1}(\partial \mathbf{X}/\partial x_{ij})]$$
$$= -(\mathbf{X}^{-1}\mathbf{A}\mathbf{X}^{-1})_{ji},$$

where the last expression on the right-hand side denotes -1 times the element in the jth row and ith column of the matrix $\mathbf{X}^{-1}\mathbf{A}\mathbf{X}^{-1}$.

Hence

$$(\partial/\partial\mathbf{X})\,\mathrm{tr}\,(\mathbf{AX}^{-1}) = -(\mathbf{X}^{-1}\mathbf{AX}^{-1})'.$$

The above partial derivatives with respect to x_{ij} and $\mathbf{X}$ have been found on the assumption that the elements of $\mathbf{X}$ are functionally independent. Suppose, however, that $\mathbf{X}$ is symmetric and that x_{ji} is treated as being identical with x_{ij}. Then the foregoing results need modification. Assuming that $\mathbf{A}$ also is symmetric, the partial derivatives with respect to x_{ij} for $i \neq j$ are given by

$$(\partial/\partial x_{ij})\,\mathrm{tr}\,(\mathbf{AX}) = 2a_{ij},$$

$$\partial|\mathbf{X}|/\partial x_{ij} = 2X_{ij},$$

$$(\partial/\partial x_{ij})\log|\mathbf{X}| = 2x^{ij}, \tag{A1.13}$$

$$(\partial/\partial x_{ij})\,\mathrm{tr}\,(\mathbf{AX}^{-1}) = -2(\mathbf{X}^{-1}\mathbf{AX}^{-1})_{ij}. \tag{A1.14}$$

If $i = j$, the factor 2 that occurs on the right hand sides of these equations is omitted.

A1.17 CONSTRUCTING AN ORTHOGONAL MATRIX WHOSE FIRST COLUMN IS GIVEN

In Chapter 6 we require the construction of an orthogonal matrix whose first column is given. We describe below a method of construction that is very simple to program for a computer.

Let $\mathbf{U}$ be the required orthogonal matrix, of order p. Denote the jth column of $\mathbf{U}$ by $\mathbf{u}_j$ and its ith element by u_{ij}. Let $\mathbf{z}_j$, for $j = 2, \ldots, p$, denote the vector of order p whose jth element is unity and whose other elements are zeros. We shall suppose, without loss of generality, that the elements of the given column $\mathbf{u}_1$ are so ordered that the first element u_{11} is not zero or near zero. This ensures that $\mathbf{u}_1$ is not linearly dependent on $\mathbf{z}_2, \ldots, \mathbf{z}_p$ or nearly so.

The vectors $\mathbf{z}_j$ enable us to construct the successive columns of $\mathbf{U}$, beginning with $\mathbf{u}_2$. Suppose that, for $1 < k \leq p$, the first $k-1$ columns already exist, and let $\mathbf{v}_k$ be the vector defined by

$$\mathbf{v}_k = \mathbf{z}_k - \sum_{j=1}^{k-1} u_{kj}\mathbf{u}_j. \tag{A1.15}$$

Then $\mathbf{v}_k$ is orthogonal to each of the first $k-1$ columns of $\mathbf{U}$ since, for $j < k$,

$$\mathbf{u}_j'\mathbf{v}_k = \mathbf{u}_j'\mathbf{z}_k - u_{kj} = 0.$$

We have also

$$\mathbf{v}_k'\mathbf{v}_k = \mathbf{v}_k'\mathbf{z}_k = v_{kk},$$

where v_{ik} denotes the ith element of $\mathbf{v}_k$. Hence the standardised vector

$$\mathbf{u}_k = (1/\sqrt{v_{kk}})\mathbf{v}_k \tag{A1.16}$$

may be taken as the kth column of $\mathbf{U}$.

Equations (A1.15) and (A1.16) enable us to find successively $\mathbf{u}_2, \ldots, \mathbf{u}_p$. For computational purposes it may be noted that, if $p \geq 3$, some of the elements of $\mathbf{U}$ are zero since, for $1 < i < k$,

$$\begin{aligned} v_{ik} &= \mathbf{v}_k'\mathbf{z}_i \\ &= \mathbf{v}_k'(\mathbf{v}_i + \sum_{j=1}^{i-1} u\,\mathbf{u}_j) = 0, \end{aligned}$$

and hence also $u_{ik} = 0$. Thus all elements of $\mathbf{U}$ above the diagonal but not in the first row are zero.

PXP

Appendix II

SOME METHODS OF MINIMISING FUNCTIONS

A2.1 A GENERAL ITERATIVE MINIMISATION PROCEDURE

In this appendix we discuss the computational problems of minimising a function of several variables by some iterative procedure with particular reference to the situations discussed in Chapters 4 and 7. The type of function that we have in mind is one related to the likelihood function of statistical theory. It is a function primarily of a number of unknown parameters but also of the observations that have been made. Accordingly we denote it here by $F(\boldsymbol{\theta}, \mathbf{X})$ where $\boldsymbol{\theta}$ denotes a vector of q parameters $\theta_1, \ldots, \theta_q$ and where $\mathbf{X}$ denotes the complete set of observations. In the problems of particular interest to us $\mathbf{X}$ may be taken as an $N \times p$ matrix whose elements represent a random sample of N observations of a vector variable $\mathbf{x}$ of order p. The minimisation of F is with respect to $\boldsymbol{\theta}$ within some parameter space Ω.

We begin by considering fairly generally iterative procedures in which a sequence of points $\boldsymbol{\theta}^{(1)}, \boldsymbol{\theta}^{(2)}, \ldots$ in the q-dimensional space Ω are constructed such that

$$F[\boldsymbol{\theta}^{(s+1)}, \mathbf{X}] < F[\boldsymbol{\theta}^{(s)}, \mathbf{X}].$$

The initial point $\boldsymbol{\theta}^{(1)}$ is usually arbitrarily chosen. In theory the procedure should continue until no further decrease in F is possible. In practice convergence is required only to a certain degree of accuracy. Hence the procedure may, for example, be terminated when the absolute magnitudes of the first-order derivatives of F with respect to the parameters are all less than some small positive value. We assume that both the first and second order derivatives of F with respect to $\boldsymbol{\theta}$ exist and are continuous at all points of Ω. We also assume that there is at least one point in Ω at which F is minimised with $\partial F/\partial\boldsymbol{\theta} = \mathbf{0}$.

The choice of the initial point $\boldsymbol{\theta}^{(1)}$ affects the number of iterations required for the procedure to terminate. As a rule the closer $\boldsymbol{\theta}^{(1)}$ is to the final point $\hat{\boldsymbol{\theta}}$ the smaller the number of iterations. In some cases F has more than one local minimum in Ω. If so, the various minima can be discovered only by starting with a succession of different initial points.

Associated with each point, or vector, $\boldsymbol{\theta}^{(s)}$ there is a gradient vector

$$\mathbf{g}^{(s)} = (\partial F/\partial \boldsymbol{\theta})_{\boldsymbol{\theta}=\boldsymbol{\theta}^{(s)}}. \tag{A2.1}$$

Let us assume that in each iteration we can by some rule construct a positive definite symmetric matrix $\mathbf{E}^{(s)}$, of order q. Then to determine the next point $\boldsymbol{\theta}^{(s+1)}$ we move or search along the direction $\mathbf{d}^{(s)}$ defined by

$$\mathbf{d}^{(s)} = -\mathbf{E}^{(s)}\mathbf{g}^{(s)}. \tag{A2.2}$$

Along this direction F may be regarded as a function $f(\alpha)$ of the distance α from the point $\boldsymbol{\theta}^{(s)}$. Thus

$$f(\alpha) = F[\boldsymbol{\theta}^{(s)}+\alpha\mathbf{d}^{(s)}, \mathbf{X}], \tag{A2.3}$$

where $\alpha \geq 0$. In practical computations it is usually convenient to scale the vector $\mathbf{d}^{(s)}$ to unit length, but for simplicity of description we shall here suppose that $\mathbf{d}^{(s)}$ remains unscaled.

Let $h(\alpha)$ denote the slope, i.e. the derivative of $f(\alpha)$. Then

$$h(\alpha) = f'(\alpha) = \mathbf{d}^{(s)\prime}\mathbf{g}_{\alpha}, \tag{A2.4}$$

where $\mathbf{g}_{\alpha}$ is the vector $\partial F/\partial \boldsymbol{\theta}$ evaluated at the point $\boldsymbol{\theta} = \boldsymbol{\theta}^{(s)}+\alpha\mathbf{d}^{(s)}$. In particular, since $\mathbf{g}_0 = \mathbf{g}^{(s)}$, the slope at $\alpha = 0$ is

$$h(0) = \mathbf{d}^{(s)\prime}\mathbf{g}^{(s)} = -\mathbf{g}^{(s)\prime}\mathbf{E}^{(s)}\mathbf{g}^{(s)}, \tag{A2.5}$$

using equation (A2.2). As $\mathbf{E}^{(s)}$ is positive definite, this is negative unless $\mathbf{g}^{(s)}$ is null, when it is zero. If $\mathbf{g}^{(s)}$ is null, the required minimum of F is located at $\boldsymbol{\theta} = \boldsymbol{\theta}^{(s)}$. Otherwise $f(\alpha)$ has a minimum for some $\alpha > 0$. In the latter case suppose that we can by some means find an approximation $\alpha^{(s)}$ to the minimising value of α. Then the new point $\boldsymbol{\theta}^{(s+1)}$ is defined by

$$\boldsymbol{\theta}^{(s+1)} = \boldsymbol{\theta}^{(s)}+\alpha^{(s)}\mathbf{d}^{(s)}. \tag{A2.6}$$

In the following sections we consider, in effect, various methods of determining $\mathbf{E}^{(s)}$ and $\alpha^{(s)}$.

A2.2 METHOD OF STEEPEST DESCENTS

In the method of steepest descents the matrix $\mathbf{E}^{(s)}$ is at each iteration taken to be simply the unit matrix. In that case $\mathbf{d}^{(s)} = -\mathbf{g}^{(s)}$. This is the direction for which the value of $-h(0)$ is maximised and along which the function value in the neighbourhood of $\boldsymbol{\theta}^{(s)}$ is decreasing most rapidly. A trial value α^* is chosen, often in some arbitrary manner, and the function and slope values at $\alpha = 0$ and $\alpha = \alpha^*$ are calculated. Interpolation or extrapolation is then used to determine $\alpha^{(s)}$. If $h(\alpha^*)$ is found to be positive, cubic interpolation for the value of α that minimises $f(\alpha)$, using the values of $f(0)$, $f(\alpha^*)$, $h(0)$ and $h(\alpha^*)$, is usually effective. If, on the other hand, $h(\alpha^*)$ is negative but smaller in absolute magnitude than $h(0)$, linear extrapolation for the zero of $h(\alpha)$, using only the values of $h(0)$ and $h(\alpha^*)$ generally works satisfactorily.

Sometimes, when the value of α^* happens to have been badly chosen, the above method breaks down. In that case a more elaborate search procedure for $\alpha^{(s)}$ is required, but we shall not concern ourselves with the details of this. The interpolated or extrapolated value of α may in practice be considered satisfactory if $|h(\alpha)| \leq 0{\cdot}1|h(0)|$. If this condition is not satisfied, further interpolation or extrapolation may be carried out.

The method of steepest descents usually works well if the point $\boldsymbol{\theta}^{(s)}$ is far from the minimum point $\hat{\boldsymbol{\theta}}$. It is much less effective when $\boldsymbol{\theta}^{(s)}$ is close to $\hat{\boldsymbol{\theta}}$. For this reason it is best to use the method only in the first few iterations and then to continue with one of the methods described subsequently.

A2.3 THE NEWTON–RAPHSON METHOD

Let $\mathbf{G}$ be the matrix $\partial^2 F/\partial\boldsymbol{\theta}\,\partial\boldsymbol{\theta}'$, symmetric and of order q, of which the element in the ith row and jth column is the second-order derivative $\partial^2 F/\partial\theta_i\partial\theta_j$. Let us suppose that $\boldsymbol{\theta}^{(s)}$ belongs to a neighbourhood of the minimum point $\hat{\boldsymbol{\theta}}$ within which F is approximately a quadratic function of the parameters and $\mathbf{G}$ is positive definite. Then, since $\partial F/\partial\boldsymbol{\theta}$ is zero at $\boldsymbol{\theta} = \hat{\boldsymbol{\theta}}$, we have approximately,

$$-\mathbf{g}^{(s)} = \mathbf{G}^{(s)}[\hat{\boldsymbol{\theta}} - \boldsymbol{\theta}^{(s)}],$$

where $\mathbf{G}^{(s)}$ denotes $\mathbf{G}$ evaluated at $\boldsymbol{\theta} = \boldsymbol{\theta}^{(s)}$. This is equivalent to

$$\hat{\boldsymbol{\theta}} \approx \boldsymbol{\theta}^{(s)} - \mathbf{E}^{(s)}\mathbf{g}^{(s)},$$

where $\mathbf{E}^{(s)}$ is the inverse of $\mathbf{G}^{(s)}$, which exists and is positive definite.

The above approximation suggests that we define $\boldsymbol{\theta}^{(s+1)}$ as

$$\begin{aligned}\boldsymbol{\theta}^{(s+1)} &= \boldsymbol{\theta}^{(s)} - \mathbf{E}^{(s)}\mathbf{g}^{(s)} \\ &= \boldsymbol{\theta}^{(s)} + \mathbf{d}^{(s)},\end{aligned}$$

which is identical with equation (A2.6) if we take $\alpha^{(s)} = 1$. This constitutes the Newton–Raphson method. It is not in fact necessary to find the inverse matrix $\mathbf{E}^{(s)}$, since $\mathbf{d}^{(s)}$ may be obtained as the solution of the equation

$$\mathbf{G}^{(s)}\mathbf{d}^{(s)} = -\mathbf{g}^{(s)}.$$

In practice, unless $\boldsymbol{\theta}^{(s)}$ is fairly close to $\hat{\boldsymbol{\theta}}$, the approximation made above may not be very good. Some improvement in the method is then made possible by searching along the direction $\mathbf{d}^{(s)}$ instead of taking $\alpha^{(s)} = 1$ in (A2.6). The interpolation and extrapolation procedure of the preceding section may be used with $\alpha^* = 1$ as the trial value of α. As before, $\alpha^{(s)}$ is obtained as an approximation to the value of α for which $f(\alpha)$ is minimised.

The Newton–Raphson method may not work well if $\boldsymbol{\theta}^{(s)}$ is too far from $\hat{\boldsymbol{\theta}}$ and may even break down as a result of the matrix $\mathbf{G}^{(s)}$ not being positive definite. In that case it is necessary to use some other method such as that of steepest descents.

A2.4 USE OF THE INFORMATION MATRIX

In some statistical problems the second-order derivatives of F are complicated expressions whose evaluation requires lengthy computation. If the sample size N is large, these derivatives will, in a probability sense, be near to their respective expectations. In that case the matrix

$$\boldsymbol{\Gamma} = E(\mathbf{G}) = E(\partial^2 F/\partial\boldsymbol{\theta}\,\partial\boldsymbol{\theta}'),$$

of which the element in the ith row and jth column is $E(\partial^2 F/\partial\theta_i\,\partial\theta_j)$, may be used as an approximation to $\mathbf{G}$. The chief merit of this approximation is that in many cases $\boldsymbol{\Gamma}$ is much simpler to compute than $\mathbf{G}$. In the minimisation problems of Chapters 4 and 7 the function to be minimised differs from $-(2/n)\log_e L$, where $n = N-1$ and where L is the likelihood function, only by a function of the observations. Hence

$$E(-\partial^2 \log L/\partial\boldsymbol{\theta}\,\partial\boldsymbol{\theta}') = \tfrac{1}{2}n\boldsymbol{\Gamma}.$$

This is known in estimation theory as the information matrix.

Let $\boldsymbol{\Gamma}^{(s)}$ denote $\boldsymbol{\Gamma}$ evaluated at $\boldsymbol{\theta} = \boldsymbol{\theta}^{(s)}$. Then, provided that $\boldsymbol{\Gamma}^{(s)}$ is positive definite, we may use the method of the preceding section with $\boldsymbol{\Gamma}^{(s)}$ in place of $\mathbf{G}^{(s)}$. Unless the number of parameters is fairly small this method usually converges much more slowly than the Newton–Raphson method, as in practice $\boldsymbol{\Gamma}^{(s)}$ is seldom a very close approximation to $\mathbf{G}^{(s)}$. The method is as a rule adequate, however, if a very accurate solution is not required. The larger number of iterations needed for convergence is to some extent compensated by the smaller amount of computation carried out in each of them.

A2.5 THE FLETCHER AND POWELL METHOD

When the number of parameters is large the Newton–Raphson method, or even that of the preceding section, may require a considerable amount of computation, since the matrix **G**, or **Γ**, must be evaluated in each iteration. Here we describe a method that avoids this difficulty. The basic idea of the method is due to Davidon (1959), but it was given further development by Fletcher and Powell (1963), who provided an elegant theoretical basis for it and proofs of stability and convergence. The Fletcher and Powell method makes use of an initial **E** matrix, which is modified in a simple manner at each iteration. If the iterative procedure is continued long enough, the final **E** matrix is close to the inverse of **G** evaluated at $\boldsymbol{\theta} = \hat{\boldsymbol{\theta}}$. Convergence is slower than with the Newton–Raphson method, but the iterations are performed more rapidly.

Let $s = 1$ now refer to the first iteration in which the Fletcher and Powell method is used. For the initial **E** matrix we could, lacking anything better, take simply $\mathbf{E}^{(1)} = \mathbf{I}$. A much better choice, which reduces considerably the number of iterations required, is to use the inverse of either $\mathbf{G}^{(1)}$ or $\boldsymbol{\Gamma}^{(1)}$, supposing that this is available and positive definite. In the minimisation problems of Chapters 4 and 7 we have taken $\mathbf{E}^{(1)}$ to be the inverse of $\boldsymbol{\Gamma}^{(1)}$.

For $s = 1, 2, \ldots$ the following method is used to obtain $\mathbf{E}^{(s+1)}$ from $\mathbf{E}^{(s)}$. Having determined $\mathbf{d}^{(s)}$, $\alpha^{(s)}$ and $\boldsymbol{\theta}^{(s+1)}$ we first compute the difference vectors

$$\mathbf{y}^{(s)} = \boldsymbol{\theta}^{(s+1)} - \boldsymbol{\theta}^{(s)}$$

and

$$\mathbf{h}^{(s)} = \mathbf{g}^{(s+1)} - \mathbf{g}^{(s)}.$$

Next we compute the vector

$$\mathbf{z}^{(s)} = \mathbf{E}^{(s)}\mathbf{h}^{(s)}$$

and the two scalars

$$\beta^{(s)} = \mathbf{y}^{(s)\prime}\mathbf{h}^{(s)}$$

and

$$\gamma^{(s)} = \mathbf{z}^{(s)\prime}\mathbf{h}^{(s)}.$$

The new **E** matrix is then given by

$$\mathbf{E}^{(s+1)} = \mathbf{E}^{(s)} + [1/\beta^{(s)}]\mathbf{y}^{(s)}\mathbf{y}^{(s)\prime} - [1/\gamma^{(s)}]\mathbf{z}^{(s)}\mathbf{z}^{(s)\prime}.$$

Note that this satisfies the equation

$$\mathbf{E}^{(s+1)}\mathbf{h}^{(s)} = \mathbf{y}^{(s)}.$$

If $\mathbf{E}^{(s)}$ is positive definite, then so is $\mathbf{E}^{(s+1)}$ provided that $\alpha^{(s)}$ is sufficiently near to the true minimum of $f(\alpha)$. As in section A2.3 the best method of choosing the trial value of α in each iteration is probably to take $\alpha^* = 1$. Though simple to use, the above method of modifying the matrix $\mathbf{E}$ is very powerful.

A2.6 MINIMISATION WITH PARAMETERS OF RESTRICTED RANGE

So far we have assumed that the q parameters all have unlimited ranges and that the parameter space is unbounded. In many problems this is not the case and some or all of the parameters may have restricted ranges. In factor analysis, for example, the residual variances ψ_i must all be non-negative. Since there is some difficulty in defining the function F when any of the ψ_i is zero, we have in fact imposed the restriction that each $\psi_i \geq \varepsilon$, where ε is some small positive number. We then require the minimisation of F within a bounded parameter space Ω whose boundary consists of all points at which one or more of the ψ_i attains the value ε.

In cases of this kind there may be no point $\boldsymbol{\theta}$ in Ω at which F has a true minimum in the sense that $\partial F/\partial\boldsymbol{\theta} = \mathbf{0}$. What then happens in the course of the iterative procedure is that, for some value of s, $\boldsymbol{\theta}^{(s)}$ lies within Ω but $\boldsymbol{\theta}^{(s+1)}$, as defined by equation (A2.6), does not. To cope with this difficulty let α_{max} denote the largest value of α for which $\boldsymbol{\theta}^{(s)}+\alpha\mathbf{d}^{(s)}$ lies within Ω. To define $\boldsymbol{\theta}^{(s+1)}$ we then substitute α_{max} for $\alpha^{(s)}$ in (A2.6). Thus $\boldsymbol{\theta}^{(s+1)}$ lies on the boundary of Ω with at least one of the restricted parameters, in our case the ψ_i, attaining its terminal value.

Suppose that the jth element of $\boldsymbol{\theta}^{(s+1)}$ attains its terminal value. Then in the $(s+1)$th and subsequent iterations the jth element of the gradient vector is set equal to zero. If the Newton–Raphson method or that of section A2.4 is in use, all non-diagonal elements of the jth row and column of $\mathbf{G}$, or of $\boldsymbol{\Gamma}$, are also set equal to zero. With the Fletcher and Powell method the non-diagonal elements of the jth row and column of $\mathbf{E}^{(s+1)}$ are set equal to zero. This has the effect of fixing the jth element of $\boldsymbol{\theta}$. Thus whenever a parameter attains its terminal value the minimisation procedure continues as if this parameter were no longer present.

The solution eventually obtained in this manner is such that all the first derivatives of F with respect to the parameters are zero except as a rule for those that have been held fixed. In the terminology of Chapter 4 the solution is improper. If any non-zero derivatives were negative, the function F could be further decreased by increasing the corresponding parameters, so a further series of iterations would be carried out. This seldom happens in practice. Usually all non-zero first order derivatives are positive. No further decrease in F is then possible without going outside Ω and the iterative procedure terminates.

As has been noted in Chapter 4, in connection with unrestricted maximum likelihood factor analysis, an improper solution is often not unique and other, entirely different, improper solutions may be discovered by starting with various different initial points.

A2.7 USE OF THE NEWTON–RAPHSON METHOD IN FACTOR ANALYSIS

In Chapter 4 the minimisation of a function f with respect to the p parameters $\psi_1, \ldots, \psi_p$ is required. As mentioned in section 4.7, exact expressions for the second-order derivatives of f have been obtained by Clarke (1970) and have been used by him in applying the Newton–Raphson method. In the notation of Chapter 4 his result is

$$\partial^2 f/\partial\psi_i\partial\psi_j = \phi_{ij}^2 - 2\delta_{ij}\psi_i^{-1}(\partial f/\partial\psi_i)$$
$$-2(\psi_i\psi_j)^{-1}\sum_{r=1}^{k}[\theta_r\omega_{ir}\omega_{jr}\sum_{m=k+1}^{p}(\theta_m-1)(\theta_r-\theta_m)^{-1}\omega_{im}\omega_{jm}]. \quad \text{(A2.7)}$$

For $\partial f/\partial\psi_i$ the expression of (4.20) may be substituted.

The expression in (A2.7) is unsatisfactory from a computational point of view since it involves all p latent roots and vectors of $\mathbf{S}^* = \mathbf{\Psi}^{-1/2}\mathbf{S}\mathbf{\Psi}^{-1/2}$. With a little algebra it may, however, be put into a more convenient form. Let $\mathbf{Z}$ be the symmetric $p\times p$ matrix of which the element in the ith row and jth column is given by

$$z_{ij} = \sum_{r=1}^{k}(\theta_r-1)\omega_{ir}\omega_{jr}+\delta_{ij}-s_{ij}/\sqrt{(\psi_i\psi_j)}.$$

For $h = 1, 2, \ldots$, let $z_{ij}^{(h)}$ denote the element in the ith row and jth column of the matrix $\mathbf{Z}^h$ (making $z_{ij}^{(1)}$ identical with z_{ij}). Define also $y_{ij}^{(h)}$ by

$$y_{ij}^{(h)} = \sum_{r=1}^{k}\theta_r(\theta_r-1)^{-h}\omega_{ir}\omega_{jr}.$$

Then an alternative expression for $\partial^2 f/\partial\psi_i\,\partial\psi_j$ that involves only the first k latent roots and vectors of $\mathbf{S}^*$ is

$$\phi_{ij}^2-2\delta_{ij}\psi_i^{-2}z_{ii}+2(\psi_i\psi_j)^{-1}\sum_{h=1}^{\infty}(-1)^{h+1}y_{ij}^{(h)}z_{ij}^{(h)}. \qquad \text{(A2.8)}$$

The infinite series on the right-hand side as a rule converges fairly rapidly, so that in practice only the first few terms are required. Inclusion of the first two should usually provide a good approximation. This ignores quantities of order $(\theta_m-1)^3$, for $m=k+1, \ldots, p$.

When the Newton–Raphson method is employed, using either equation (A2.7) or (A2.8), it is sometimes found in earlier iterations that the matrix of second-order derivatives is not positive definite. The crude approximation ϕ_{ij}^2 is then substituted for $\partial^2 f/\partial\psi_i\,\partial\psi_j$. The resulting matrix is always positive definite, as was shown in section 4.5.

It is natural to consider also the application of the Newton–Raphson method to the minimisation problems of Chapter 7. For this we need exact expressions for the second-order derivatives of the function F with respect to all q free parameters. Unfortunately these are very complicated in form. As an example let us consider $\partial^2F/\partial\lambda_{ir}\,\partial\lambda_{js}$, where λ_{ir} and λ_{js} are both assumed to be free parameters.

From equation (7.10) we have

$$\tfrac{1}{2}\partial F/\partial\lambda_{ir}=[\mathbf{\Sigma}^{-1}(\mathbf{\Sigma}-\mathbf{S})\mathbf{\Sigma}^{-1}\mathbf{\Lambda\Phi}]_{ir}.$$

Differentiation of this with respect to λ_{js}, using equation (A1.10) of Appendix I, gives

$$\begin{aligned}\tfrac{1}{2}\partial^2F/\partial\lambda_{ir}\partial\lambda_{js} &= [\mathbf{\Sigma}^{-1}(\partial\mathbf{\Sigma}/\partial\lambda_{js})\mathbf{\Sigma}^{-1}\mathbf{\Lambda\Phi}]_{ir}+[\mathbf{\Sigma}^{-1}(\mathbf{\Sigma}-\mathbf{S})\mathbf{\Sigma}^{-1}]_{ij}\phi_{rs}\\ &\quad-[\mathbf{\Sigma}^{-1}(\partial\mathbf{\Sigma}/\partial\lambda_{js})\mathbf{\Sigma}^{-1}(\mathbf{\Sigma}-\mathbf{S})\mathbf{\Sigma}^{-1}\mathbf{\Lambda\Phi}]_{ir}\\ &\quad-[\mathbf{\Sigma}^{-1}(\mathbf{\Sigma}-\mathbf{S})\mathbf{\Sigma}^{-1}(\partial\mathbf{\Sigma}/\partial\lambda_{js})\mathbf{\Sigma}^{-1}\mathbf{\Lambda\Phi}]_{ir}.\end{aligned}$$

On substituting for $\partial\mathbf{\Sigma}/\partial\lambda_{js}$ in the same way as in section 7.2, when passing from equation (7.11) to (7.12), we find, with some rearrangement of terms, that

$$\begin{aligned}\tfrac{1}{2}\partial^2F/\partial\lambda_{ir}\partial\lambda_{js} &= \sigma^{ij}(\mu_{rs}-w_{rs})+\eta_{is}\eta_{jr}-\eta_{is}v_{jr}-\eta_{jr}v_{is}\\ &\quad+(\phi_{rs}-\mu_{rs})u_{ij},\end{aligned} \qquad \text{(A2.9)}$$

where

$$\begin{aligned}u_{ij} &= [\mathbf{\Sigma}^{-1}(\mathbf{\Sigma}-\mathbf{S})\mathbf{\Sigma}^{-1}]_{ij},\\ v_{jr} &= [\mathbf{\Sigma}^{-1}(\mathbf{\Sigma}-\mathbf{S})\mathbf{\Sigma}^{-1}\mathbf{\Lambda\Phi}]_{jr},\\ w_{rs} &= [\mathbf{\Phi\Lambda}'\mathbf{\Sigma}^{-1}(\mathbf{\Sigma}-\mathbf{S})\mathbf{\Sigma}^{-1}\mathbf{\Lambda\Phi}]_{rs},\end{aligned}$$

and where η_{jr} and μ_{rs} are as defined in section 7.2.

The other second-order derivatives may be found in a similar manner. It is clear that in each iteration a considerable amount of computation would be required to evaluate them all. Hence in this case it does not seem feasible to employ the Newton–Raphson method except possibly when the number of free parameters is fairly small.

REFERENCES

AITKEN, A. C. (1934). 'Note on selection from a multivariate normal population', *Proc. Edinb. Math. Soc.*, **4,** 106–110

ANDERSON, T. W. (1958). *Introduction to Multivariate Statistical Analysis,* Wiley, New York

ANDERSON, T. W. and RUBIN, H. (1956). 'Statistical inference in factor analysis', *Proc. Third Berkeley Symp. math. Statist. Probab.*, **5,** 111–150

BARGMANN, R. (1957). *A Study of Independence and Dependence in Multivariate Normal Analysis,* University of North Carolina, Institute of Statistics Mimeo Series No. 186

BARTLETT, M. S. (1937). 'The statistical conception of mental factors', *Br. J. Psychol.*, **28,** 97–104

BARTLETT, M. S. (1938). 'Methods of estimating mental factors', *Nature, Lond.*, **141,** 609–610

BARTLETT, M. S. (1951). 'The effect of standardisation on an approximation in factor analysis', *Biometrika,* **38,** 337–344

BARTLETT, M. S. (1953). 'Factor analysis in psychology as a statistician sees it', in *Uppsala Symposium on Psychological Factor Analysis,* Nordisk Psykologi's Monograph Series No. 3, 23–24, Copenhagen: Ejnar Mundsgaards; Stockholm: Almqvist and Wiksell

BARTLETT, M. S. (1954). 'A note on the multiplying factor for various χ^2 approximations', *J. R. statist. Soc.*, **B 16,** 296–298

BOX, G. E. P. (1949). 'A general distribution theory for a class of likelihood criteria', *Biometrika,* **36,** 317–346

BROWNE, M. W. (1965). *A Comparison of Factor Analytic Techniques,* Master's Thesis, University of Witwatersrand, South Africa

BURT, C. (1949). 'Alternative methods of factor analysis and their relations to Pearson's method of principal axes', *Br. J. Psychol. statist. Sect.*, **2,** 98–121

CARROLL, J. B. (1953). 'An analytical solution for approximating simple structure in factor analysis', *Psychometrika,* **18,** 23–38

CLARKE, M. R. B. (1970). 'A rapidly convergent method for maximum likelihood factor analysis', *Br. J. math. & statist. Psychol.*, **23,** 43–52

DAVIDON, W. C. (1959). *Variable Metric Method for Minimization,* A.E.C. Research and Development Report, ANL-5990 (Rev.)

EMMETT, W. G. (1949). 'Factor analysis by Lawley's method of maximum likelihood', *Br. J. Psychol. statist. Sect.*, **2,** 90–97

FLETCHER, R. and POWELL, M. J. D. (1963). 'A rapidly convergent descent method for minimization', *Comput. J.*, **2,** 163–168

GIRSHICK, M. A. (1939). 'On the sampling theory of roots of determinantal equations', *Ann. math. Statist.*, **10,** 203–224

HENDRICKSON, A. E. and WHITE, P. O. (1964). 'Promax: a quick method for rotation to oblique simple structure', *Br. J. statist. Psychol.*, **17**, 65–70

HOLZINGER, K. J. and SWINEFORD, F. (1939). *A Study in Factor Analysis: The Stability of a Bi-factor Solution*, University of Chicago, Supplementary Educational Monographs, No. 48

HORST, P. (1965). *Factor Analysis of Data Matrices*, Holt, Rinehart and Winston, New York

HOTELLING, H. (1933). 'Analysis of a complex of statistical variables into principal components', *J. educ. Psychol.*, **24**, 417–441, 498–520

HOTELLING, H. (1957). 'The relation of the newer multivariate statistical methods to factor analysis', *Br. J. statist. Psychol.*, **10**, 69–79

HOWE, W. G. (1955). *Some Contributions to Factor Analysis*, USAEC Rep. ORNL-1919

JÖRESKOG, K. G. (1963). *Statistical Estimation in Factor Analysis*, Almqvist and Wiksell, Stockholm

JÖRESKOG, K. G. (1966a). 'UMFLA—A computer program for unrestricted maximum likelihood factor analysis', *Research Memorandum* 66–20, Princeton, N.J., Educational Testing Service

JÖRESKOG, K. G. (1966b). 'Testing a simple structure hypothesis in factor analysis', *Psychometrika*, **31**, 165–178

JÖRESKOG, K. G. (1967). 'Some contributions to maximum likelihood factor analysis', *Psychometrika*, **32**, 443–482

JÖRESKOG, K. G. (1969). 'A general approach to confirmatory maximum likelihood factor analysis', *Psychometrika*, **34**, 183–202

JÖRESKOG, K. G. and GRUVAEUS, G. (1967). 'RMLFA—A computer program for restricted maximum likelihood factor analysis', *Research Memorandum* 67–21, Princeton, N.J., Educational Testing Service

JÖRESKOG, K. G. and LAWLEY, D. N. (1968). 'New methods in maximum likelihood factor analysis', *Br. Jnl math. &. statist. Psychol.*, **21**, 85–96

KAISER, H. F. (1958). 'The varimax criterion for analytic rotation in factor analysis', *Psychometrika*, **23**, 187–200

KAISER, H. F. and DICKMAN, K. W. (1959). *Analytical Determination of Common Factors*, unpublished Manuscript

LAWLEY, D. N. (1940). 'The estimation of factor loadings by the method of maximum likelihood', *Proc. R. Soc. Edinb.* **A 60,** 64–82

LAWLEY, D. N. (1941). 'Further investigations in factor estimation', *Proc. R. Soc. Edinb.* **A 61,** 176–185

LAWLEY, D. N. (1943). 'The application of the maximum likelihood method to factor analysis', *Br. J. Psychol.*, **33,** 172–175

LAWLEY, D. N. (1953). 'A modified method of estimation in factor analysis and some large sample results', *Uppsala Symposium on Psychological Factor Analysis*, Nordisk Psykologi's Monograph Series No. 3, 35–42, Copenhagen: Ejnar Mundsgaards; Stockholm: Almqvist and Wiksell

LAWLEY, D. N. (1956). 'Tests of significance for the latent roots of covariance and correlation matrices', *Biometrika*, **43,** 128–136

LAWLEY, D. N. (1958). 'Estimation in factor analysis under various initial assumptions', *Br. J. statist. Psychol.*, **11,** 1–12

LAWLEY, D. N. (1963). 'On testing a set of correlation coefficients for equality', *Ann. math. statist.*, **34,** 149–151

LAWLEY, D. N. (1967). 'Some new results in maximum likelihood factor analysis, *Proc. R. Soc. Edinb.*, **A 67,** 256–264

LAWLEY, D. N. and MAXWELL, A. E. (1964). 'Factor transformation methods', *Br. J. statist. Psychol.*, **17,** 97–103

McDonald, R. P. (1967a). 'Numerical methods for polynomial models in non-linear factor analysis', *Psychometrika*, **32**, 77–112

McDonald, R. P. (1967b). 'Factor interaction in non-linear factor analysis', *Br. Jnl math. & statist. Psychol.*, **20**, 215

Mattsson, A., Olsson, U. and Rosén, M. (1966). '*The Maximum Likelihood Method in Factor Analysis with Special Consideration to the Problem of Improper Solutions*, Research Report, Institute of Statistics, University of Uppsala, Sweden

Maxwell, A. E. (1961). 'Recent trends in factor analysis', *J. R. statist. Soc.*, **A 124**, 49–59

Pearson, K. (1901). 'On lines and planes of closest fit to a system of points in space', *Phil. Mag.*, **2**, 6th Series, 557–572

Pearson, K. (1912). 'On the general theory of the influence of selection on correlation and variation', *Biometrika*, **8**, 437–443

Spearman, C. (1904). 'General intelligence objectively determined and measured', *Am. J. Psychol.*, **15**, 201–293

Spearman, C. (1926). *The Abilities of Man*, MacMillan, London

Thomson, G. H. (1951). *The Factorial Analysis of Human Ability*, London University Press

Thurstone, L. L. (1947). *Multiple Factor Analysis*, University of Chicago Press

Tumura, Y., Fukutomi, K. and Assoo, Y. (1968). 'On the unique convergence of iterative procedures in factor analysis', *TRU Math.*, **4**, 52–59

Whittle, P. (1953). 'A principal component and least squares method of factor analysis', *Skand. Aktuarietidskrift*, **35**, 223–239

Wilkinson, J. H. (1965). *The Algebraic Eigenvalue Problem*, Clarendon Press, Oxford

INDEX